Introduction to
Complex Variables

Merrill Mathematics Series

Erwin Kleinfeld, Editor

Introduction to Complex Variables

Peter Colwell
Jerold C. Mathews
Iowa State University

Charles E. Merrill Publishing
A Bell & Howell Company
Columbus, Ohio 43216

69438

Published by
Charles E. Merrill Publishing Co.
A Bell & Howell Company
Columbus, Ohio 43216

International Standard Book Number: 0–675–08974–3

Library of Congress Catalog Card Number: 72–95500

AMS 1971 Subject Classification 3001

1 2 3 4 5 6 7 8 9 10 — 78 77 76 75 74 73

Printed in the United States of America

Preface

The uses and users of complex analysis appear in all the physical and engineering sciences. This book offers a brief, efficient introduction to the subject which develops most of the techniques we perceive to be basic and common to all users. Most modern applications of complex analysis require more than these basic techniques. For this reason an introduction to complex variable methods must try to show why the techniques work, as well as what they are. We have chosen to do this by stating the facts as theorems whose hypotheses and conclusions are explicitly listed and by including those proofs which help explain the facts. Our development will help most users to acquire the basic skills and prepare them to understand additional techniques of more special interest to them. And it will let students of mathematics see the operational facts emerge from the elementary analysis.

We have some specific comments for the reader. "Elementary" does not mean "easy." To follow our development, you must come prepared. We assume you are ready to use the tools of calculus of one and several variables, line integrals, infinite series and power series. To master the techniques, you must do the problems. We specifically invite those who use this book for self-study to write us if they have questions.

A professional staff at Charles E. Merrill has given us much help. We particularly wish to thank Pete Hutton and Linda Mohrman, the Series Editor and our personal taskmaster.

Ames, Iowa
P.C.
J.C.M.

Contents

Functions of a Complex Variable

In some ways the beginning of a study of complex-valued functions of a complex variable is very much like an elementary course in calculus, primarily because these functions include all the real-valued functions of a real variable which were studied in calculus. The elementary calculus gives us a natural vocabulary to use and a natural sequence of ideas and operations to follow in developing the "calculus of complex-valued functions." This natural sequence is really the outline of this chapter, and so there will be nothing in it which we can say is really new or surprising.

Section 1.1   Complex Numbers

Since the numbers we deal with are to be complex numbers, we first review the usual terminology and operations with complex numbers.

The complex plane is the familiar two-dimensional Cartesian plane, R^2, whose points are the complex numbers, z, identified by specifying a pair of real numbers x and y. We write $z = x + iy$, where i satisfies the equation $i^2 = -1$. We can think of z as the point (x,y) in R^2 or as the two-dimensional vector (x,y) over the reals.

For $z = x + iy$, we call $x = \text{Re } z$ the *real part* of z, $y = \text{Im } z$ the *imaginary part* of z, $\bar{z} = x - iy$ the *conjugate* of z, and $|z| = (x^2 + y^2)^{1/2}$ the *modulus*, or *absolute value*, of z. Note that $|z|^2 = z\bar{z}$, while $|\text{Re } z| \leq |z|$, $|\text{Im } z| \leq |z|$, and $|z| \leq |\text{Re } z| + |\text{Im } z|$.

If we represent R^2 with polar coordinates, and (r,θ) is a representation in polar coordinates for the point (x,y) of R^2 where $(x,y) \neq (0,0)$, then for $z = x + iy$ we have z in *polar form* given by $z = r(\cos \theta + i \sin \theta)$, where $r = |z| \neq 0$. We call $\theta = \text{arg } z$ *an argument of* z. There are infinitely many values of θ for which (r,θ) in polar coordinates describes the point (x,y) in rectangular coordinates, so we can expect z to have infinitely many arguments. However, z will have only one argument θ for which $-\pi < \theta \leq \pi$, and it is this argument we call the *principal argument of* z. We denote it by $\text{Arg } z$. This choice for a principal argument of z assigns principal argument 0 to each positive real number. We do not define any argument for the complex number 0. One of the advantages of the polar form is that with it we can conveniently obtain positive integral powers of a complex number through the use of *De-Moivre's Theorem:* If $z = r(\cos \theta + i \sin \theta)$ and n is any positive integer, then

$$z^n = r^n(\cos n\theta + i \sin n\theta).$$

Let $z_1 = x_1 + iy_1 = r_1(\cos \theta_1 + i \sin \theta_1)$ and $z_2 = x_2 + iy_2 = r_2(\cos \theta_2 + i \sin \theta_2)$ be two complex numbers. We give the usual algebraic operations on complex numbers.

$$z_1 + z_2 = (x_1 + x_2) + i(y_1 + y_2)$$

$$z_1 z_2 = (x_1 y_1 - x_2 y_2) + i(x_1 y_2 + x_2 y_1)$$

$$= r_1 r_2 [\cos(\theta_1 + \theta_2) + i \sin(\theta_1 + \theta_2)]$$

$$\frac{z_1}{z_2} = \frac{[(x_1 x_2 + y_1 y_2) + i(-x_1 y_2 + x_2 y_1)]}{(x_2^2 + y_2^2)}$$

$$= \left(\frac{r_1}{r_2}\right)[\cos(\theta_1 - \theta_2) + i \sin(\theta_1 - \theta_2)]$$

as long as $x_2^2 + y_2^2 = r_2^2 \neq 0$.

If $r_2 \neq 0$, we can also write

$$\frac{z_1}{z_2} = \frac{z_1 \bar{z}_2}{|z_2|^2} = \frac{z_1 \bar{z}_2}{r_2^2}.$$

In particular,

$$\frac{1}{z_2} = \frac{(x_2 - iy_2)}{(x_2^2 + y_2^2)}$$

$$= \left(\frac{1}{r_2}\right)(\cos \theta_2 - i \sin \theta_2).$$

The operations of multiplication and division are the only items we have mentioned so far which distinguish the complex number $z = x + iy$ from the vector (x,y).

Example 1.1   Suppose $z_1 = i$ and $z_2 = 1 - i$. In polar form $z_1 = 1[\cos (\pi/2) + i \sin (\pi/2)]$, and $z_2 = 2^{1/2}[\cos (-\pi/4) + i \sin (-\pi/4)]$.

$$z_1 z_2 = 1 + i = 2^{1/2}\left[\cos\left(\frac{\pi}{4}\right) + i \sin\left(\frac{\pi}{4}\right)\right]$$

$$\frac{z_1}{z_2} = \frac{z_1 \bar{z}_2}{|z_2|^2} = \frac{(-1 + i)}{2}$$

$$= 2^{-1/2}\left[\cos\left(\frac{3\pi}{4}\right) + i \sin\left(\frac{3\pi}{4}\right)\right]$$

$$\frac{1}{z_2} = 2^{-1/2}\left[\cos\left(\frac{\pi}{4}\right) + i \sin\left(\frac{\pi}{4}\right)\right].$$

Example 1.2   Find all the solutions of the equation $z^4 = 1$. If we assume that $z = r(\cos \theta + i \sin \theta)$ is a solution, then $r^4(\cos 4\theta + i \sin 4\theta) = 1$.

For any integer k we can write $1 = 1(\cos 2\pi k + i \sin 2\pi k)$. If we take $r = 1$ and $\theta = \pi k/2$ for $k = 0, 1, 2, 3$, we obtain four distinct solutions to the equation $z^4 = 1$ of the form $z = 1(\cos \pi k/2 + i \sin \pi k/2)$:

$$k = 0; \quad z = 1$$

$$k = 1; \quad z = 1\left(\cos \frac{\pi}{2} + i \sin \frac{\pi}{2}\right) = i$$

$$k = 2; \quad z = 1(\cos \pi + i \sin \pi) = -1$$

$$k = 3; \quad z = 1\left(\cos \frac{3\pi}{2} + i \sin \frac{3\pi}{2}\right) = -i$$

The reader should verify that for any value of k other than 0, 1, 2, or 3, $z = 1(\cos \pi k/2 + i \sin \pi k/2)$ will be one of the four numbers 1, i, -1, $-i$.

We also need to have some terminology to describe those kinds of sets of complex numbers which we use frequently. If a is a complex number and r is a positive real number, the inequalities $|z - a| < r$, $|z - a| \le r$, $|z - a| = r$ will denote, respectively, the open disk of radius r with center at $z = a$, the closed disk of radius r with center at $z = a$, and the circle of radius r with center at $z = a$. More generally, a set S of complex numbers is *open* if each point of S is the center of an open disk of positive radius, each point of which belongs to S; in more formal terms, S is open if for every $z_0 \in S$ there is a number $r(z_0) > 0$ such that every point z with $|z - z_0| < r(z_0)$ lies in S.

An open set S of complex numbers is *connected* if each pair of points in S can be joined by a polygonal path lying entirely in S, and we call S a *domain* if S is open and connected. A domain S is *simply connected* if every closed curve in S contains only points of S inside it.

Example 1.3   Let $S = \{z : |z| < 1\}$ and $T = \{z : 0 < |z| < 1\}$. Both S and T are domains; S is simply connected, but T is not simply connected.

The word "domain" is more frequently used to describe the set on which a function is defined. Since we will find it convenient to define most functions we consider on open, connected sets, there will be no real ambiguity in the use of the word.

Example 1.4   Let H be the set of numbers $z = x + iy$ for which $|z - 1| < |z|$. Then $|z - 1|^2 < |z|^2$, $(z - 1)\overline{(z - 1)} < z\bar{z}$, $(z - 1)(\bar{z} - 1) < z\bar{z}$, and $1 < z + \bar{z}$. Since $z + \bar{z} = 2\,\mathrm{Re}\,z$, this means that $\mathrm{Re}\,z > \frac{1}{2}$ for every z in H.

Since it is easily checked that every z with $\mathrm{Re}\,z > \frac{1}{2}$ lies in H, we can call H a half-plane to the right of the line $\mathrm{Re}\,z = \frac{1}{2}$. Note that H is a domain.

PROBLEMS

1.1   For each number z given, find $2z$, z^3, $1/z$, $\bar{z}$, and $z/|z|$ in both rectangular and polar forms. Plot the following numbers in the Cartesian plane.
(a) $z = i$
(b) $z = 1 + i$
(c) $z = 3 + 4i$

1.2   Let $z_1 = 1 + i$, $z_2 = 2 - i$.

(a)   Write z_1 and z_2 in polar form with principal arguments.

(b)   Find $z_1 + z_2$, $z_1 z_2$, z_1/z_2 in rectangular and polar forms.

(c)   Describe each of the following sets geometrically and by giving an equation in x and y which the points of each set must satisfy:

(i) J is the set of all z for which

$$|z - z_1| = |z - z_2|.$$

(ii) K is the set of all z for which

$$|z - z_1| + |z - z_2| = 3.$$

1.3   Verify the following statements.

(a) $|z^n| = |z|^n$ for any integer n and any complex number z.

(b) $|z_1 + z_2 + \cdots + z_n| \leq |z_1| + |z_2| + \cdots + |z_n|$ for positive integer n.

1.4   What geometric condition must z_1 and z_2 satisfy if $|z_1 + z_2| = |z_1| + |z_2|$?

1.5   Show that $|z_1 + z_2| = |z_1| - |z_2|$ if and only if $\arg z_1 - \arg z_2 = k\pi$ for some odd integer k.

1.6   If a and b are real numbers, describe geometrically the set of complex numbers z for which $|(z - a)/(z - b)| > 1$.

1.7   Find all solutions to each of the following equations.

(a) $z^3 = 1$

(b) $z^5 = -1$

(c) $z^6 = 3$

(d) $z^2 + z + 3 = 0$

(e) $z^3 = 8i$

1.8   Suppose w is any solution to the equation $z^n = 1$, where n is a positive integer and $\operatorname{Im} w \neq 0$. Prove that

$$1 + w + w^2 + \cdots + w^{n-1} = 0.$$

1.9   Use DeMoivre's Theorem to show that

(a) $\cos 5x = 16 \cos^5 x - 20 \cos^3 x + 5 \cos x$ for all real x.

(b) $\sin(x+y) = \sin x \cos y + \cos x \sin y$ for all real x and y.

(c)   Find a formula for the sum

$$1 + \cos x + \cos 2x + \cdots + \cos(n-1)x$$

for all real x and any positive integer n.

1.10   Which of the following sets are open? connected? domains? simply connected domains?
(a) $A = \{z: |z-1| = 2\}$
(b) $B = \{z: |z-1| \le 2\}$
(c) $C = \{z: |z-1| < 2\}$
(d) $D = \{z: |z-1| > 2\}$
(e) $E = \{z: 1 < |z| < 2\}$
(f) $F = \{z: 1 < |z| \le 2\}$
(g) $G = \{z = x + iy: xy > 0\}$

Section 1.2   Functions and Mappings

Definition   Let D be a domain. If to every point z of D we assign a unique complex number $w = f(z)$, we say that the equation $w = f(z)$ defines a *complex-valued function* on D. We call D the *domain of the function*, and for each z in D we call $w = f(z)$ the *image* of z. The set of all images $\{w: w = f(z), z \in D\}$ is called the *range of the function*.

No ambiguity results from using $f(z)$, or the equation $w = f(z)$, or even f, to denote the function defined on D. If we let E be the range of the function $f(z)$, we shall also call the function $f(z)$ a *mapping* from domain D to the set E. For illustration let us define two particularly simple mappings for which $D = E =$ the complex plane. First, for any fixed complex number b, let $f(z) = z + b$ for any complex number z. Second, for any fixed non-zero complex number a, define $f(z) = az$ for any complex number z. The first of these functions may be described geometrically by observing that it just translates the entire complex plane a distance $|b|$ units in the direction Arg b. The second function produces a rotation of each z through the angle Arg a and changes the distance of z from 0 by a factor $|a|$.

If $w = f(z)$ is a mapping from D to E, where $z = x + iy$, $w = u + iv$, and x, y, u, v are real, we can write $f(x + iy) = u(x,y) + iv(x,y)$ and think of the mapping in terms of a pair of real-valued functions, $u(x,y)$ and $v(x,y)$, defined on the set D in R^2. For the present we want to describe the behavior of $f(z)$ as a complex-valued function as fully as we can in terms of the behavior of the real-valued functions $u(x,y) = \text{Re}\, f(z)$ and $v(x,y) = \text{Im}\, f(z)$. This means we shall be treating $f(z)$ as if it were only a vector-valued function of two real variables. The point in our discussion where we once again look at $f(z)$ as a function of a single complex variable is where we examine the notion of a derivative. After this point we shall seldom find it advantageous to think of $f(z)$ as a vector-valued function of two real variables.

Definition A function $f(z)$ defined in domain D has a *limit at z_0* in D if there exists a complex number L with the following property: For every number $\epsilon > 0$ there exists a number $\delta(\epsilon, z_0) > 0$ (depending on z_0 and ϵ) such that $|f(z) - L| < \epsilon$ whenever $z \in D$ and $0 < |z - z_0| < \delta(\epsilon, z_0)$. We call L the limit of $f(z)$ at z_0 and write $\lim_{z \to z_0} f(z) = L$.
(An equivalent definition of limit, in terms of convergence of sequences, is given in problem 1.16.)
The reader can supply a proof of the following statement.

THEOREM 1.1 Hypotheses: D is a domain, $z_0 = x_0 + iy_0 \in D$, and $f(z)$ is a function defined in D; $L = A + iB$ is a given complex number.

Conclusion: $\lim_{z \to z_0} f(z) = L$ if and only if both

$$\lim_{(x,y) \to (x_0, y_0)} \text{Re}\, f(z) = A \quad \text{and} \quad \lim_{(x,y) \to (x_0, y_0)} \text{Im}\, f(z) = B.$$

The definition and theorem above are narrower than is necessary. We need not require $f(z)$ to be defined at a point z_0 in order to be able to speak of $f(z)$ as having a limit at z_0. In both the definition and the theorem it is enough to require that $f(z)$ be defined in a domain D which for some number $r > 0$ contains the set $\{z : 0 < |z - z_0| < r\}$ about the point z_0.

Definition Let $f(z)$ be defined in a domain D and let z_0 be a point of D. Then $f(z)$ is *continuous at z_0* if $f(z)$ has limit $f(z_0)$ at z_0; $f(z)$ is *continuous in D* if $f(z)$ is continuous at every point of D.

With Theorem 1.1 it is not hard to prove this result.

THEOREM 1.2 Hypotheses: $f(z)$ is defined in a domain D and $z_0 = x_0 + iy_0 \in D$.

Conclusion: $f(z)$ is continuous at z_0 if and only if both $\text{Re}\, f(z)$ and $\text{Im}\, f(z)$ are continuous at (x_0, y_0).

There is nothing unfamiliar in the definitions and theorems above, and the usual body of properties and facts related to them are available. We summarize them in the problems below; numerous as they are, they show us that the concepts of complex-valued function, limit, and continuity are ideas we already recognize as familiar ones.

PROBLEMS

1.11 Prove that if $f(z)$ is defined in a domain D and has a limit at a point z_0 of D, then this limit is unique.

1.12 Prove Theorem 1.2.

1.13 Let $f(z)$ and $g(z)$ be defined in a domain D containing z_0, and suppose that $\lim_{z \to z_0} f(z) = L$, $\lim_{z \to z_0} g(z) = M$.
Prove that:
(a)   for any complex numbers α, β,

$$\lim_{z \to z_0}[\alpha f(z) + \beta g(z)] = \alpha L + \beta M$$

(b) $\lim_{z \to z_0} f(z) g(z) = LM$

(c) $\lim_{z \to z_0} f(z)/g(z) = L/M$ if $M \neq 0$.

1.14 If $f(z)$ and $g(z)$ are defined in a domain D and continuous at a point z_0 of D, then the following functions are also continuous at z_0:
(a) $\alpha f(z) + \beta g(z)$, where α and β are any complex numbers
(b) $f(z) g(z)$
(c) $f(z)/g(z)$ if $g(z_0) \neq 0$.

1.15 Let D be the complex plane. Decide, for each of the following functions, at which points the functions will not be continuous; will not have a limit; will have a limit but will not be continuous:
(a) $f(z) = |z|$
(b) $f(z) = |z|/z$
(c) $f(z) = |z|^2/z$
(d) $f(z) = 1/(z^2 + 1)$
(e) $f(z) = (z - i)/(z^2 + 1)$.

1.16 Let $\{a_n\}_{n=1}^{\infty} = a_1, a_2, \ldots, a_n, \ldots$ be a sequence of complex numbers. We say the sequence $\{a_n\}_{n=1}^{\infty}$ converges to A, or has limit A, and write $\lim_{n \to \infty} a_n = A$, if for every number $\epsilon > 0$ there is an integer $N(\epsilon) > 0$ such that $|a_n - A| < \epsilon$ whenever $n > N(\epsilon)$.
Suppose $f(z)$ is defined in a domain containing z_0.

(a)   Prove that $\lim_{z \to z_0} f(z) = L$ if and only if for every sequence $\{z_n\}_{n=1}^{\infty}$ with $\lim_{n \to \infty} z_n = z_0$, it is true that $\lim_{n \to \infty} f(z_n) = L$.

(b)   Prove that $f(z)$ is continuous at z_0 if and only if $\lim_{n \to \infty} f(z_n) = f(z_0)$ for every sequence $\{z_n\}_{n=1}^{\infty}$ in D with $\lim_{n \to \infty} z_n = z_0$.

1.17 A set S of complex numbers is called *bounded* if for some finite number $M > 0$, $|z| \leq M$ for every $z \in S$.

   We call S *closed* if the set of all complex numbers *not* in S is open. And we call S *compact* if S is closed and bounded. Which of the following sets are compact?

(a) $S = \{z : |z| < 1\}$
(b) $S = \{z : |z| = 1\}$
(c) $S = \{z : |z| \leq 1\}$
(d) $S = \{z : |z| < 1 \text{ and } \operatorname{Im} z > 0\}$
(e) $S = \{z : |z| \leq 1 \text{ and } \operatorname{Im} z > 0\}$
(f) $S = \{z : |z| \leq 1 \text{ and } \operatorname{Im} z \geq 0\}$.

1.18 Suppose $f(z)$ is continuous in a domain D containing a compact set M.

(a)   Prove that the range of f on M is a bounded set.

(b)   If K is the least upper bound of the numbers $|f(z)|$ for $z \in M$, prove that there is a point $z_0 \in M$ for which $|f(z_0)| = K$.

1.19 Let $f(z)$ be defined in a domain D containing a compact set M. We say $f(z)$ is *uniformly continuous on M* if for every number $\epsilon > 0$ there exists a number $\delta(\epsilon) > 0$ such that $|f(z) - f(w)| < \epsilon$ whenever $z, w \in M$ and $|z - w| < \delta(\epsilon)$.

(a)   Prove that if $f(z)$ is uniformly continuous on M then $f(z)$ is continuous on M.

(b)   Prove that if $f(z)$ is continuous on D and M is a compact set in D, then $f(z)$ is uniformly continuous on M.

(c)   Give an example to show that statement (b) is not necessarily true if the word "compact" is deleted.

Section 1.3   Derivatives and Analytic Functions

The definition we give for the derivative of a complex-valued function at a point will be very familiar in a formal sense. However, it will become apparent that there are fundamental differences between real-valued and complex-valued functions as far as derivatives are concerned.

Definition   Let $f(z)$ be defined in a domain D and z_0 be a point of D. The *derivative of $f(z)$ at z_0* is defined to be the limit

$$\lim_{h \to 0} \left[\frac{f(z_0 + h) - f(z_0)}{h} \right].$$

(Here h is a complex number.) If this limit exists, we say $f(z)$ is *differentiable at z_0*, and we write this limit as $f'(z_0)$, or $\left. \dfrac{df}{dz} \right|_{z = z_0}$. We say $f(z)$ is differentiable in D if $f(z)$ is differentiable at each point of D.

Example 1.4   The function $f(z) = z^2 = (x^2 - y^2) + i(2xy)$ is defined for all z. For any z_0 and any complex number $h \neq 0$,

$$\frac{[f(z_0 + h) - f(z_0)]}{h} = \frac{[(z_0 + h)^2 - z_0^2]}{h}$$

$$= \frac{(2z_0 h + h^2)}{h}$$

$$= 2z_0 + h,$$

so

$$f'(z_0) = \lim_{h \to 0} (2z_0 + h) = 2z_0.$$

This is exactly the result we might have expected.

Example 1.5   If we define

$$f(z) = \begin{cases} \dfrac{|z|^5}{z^4}, & z \neq 0 \\ 0, & z = 0 \end{cases},$$

then $f(z)$ is continuous at $z = 0$. For $z_0 = 0$ we see that for h real,

$$\lim_{h \to 0} \frac{[f(0 + h) - f(0)]}{h} = \lim_{h \to 0} 1 = 1;$$

but for $h = ir$ where $r > 0$,

$$\lim_{h \to 0} \frac{[f(0 + h) - f(0)]}{h} = \lim_{r \to 0} \left(\frac{r^5}{ir^5} \right) = \frac{1}{i} = -i.$$

Thus $f(z)$ does not have a derivative at $z_0 = 0$. What is the case for any $z_0 \neq 0$?

The usual properties of differentiation hold for functions of a complex variable, and the proofs are formally the same as those for real-valued functions of a real variable.

THEOREM 1.3 Hypotheses: $f(z)$ and $g(z)$ are defined in a domain D and are differentiable at $z_0 \in D$.

Conclusions:
C1 Both $f(z)$ and $g(z)$ are continuous at z_0.
C2 If for any complex numbers α and β, $F(z) = \alpha f(z) + \beta g(z)$, then $F(z)$ is differentiable at z_0, and

$$F'(z_0) = \alpha f'(z_0) + \beta g'(z_0).$$

C3 If $G(z) = f(z) g(z)$, then $G(z)$ is differentiable at z_0, and $G'(z_0) = f(z_0) g'(z_0) + f'(z_0) g(z_0)$.
C4 If $H(z) = f(z)/g(z)$, and $g(z_0) \neq 0$, then $H(z)$ is differentiable at z_0, and

$$H'(z_0) = \frac{[g(z_0)f'(z_0) - f(z_0)g'(z_0)]}{[g(z_0)]^2}.$$

C5 If $f(z) = c$ for some complex number c, then $f(z)$ is differentiable at z_0, and $f'(z_0) = 0$.

Example 1.6 Let $f(z) = |z|^2$ for all z. We shall show that $f(z)$ is differentiable only at $z_0 = 0$. Now

$$f'(0) = \lim_{h \to 0} \frac{[f(0 + h) - f(0)]}{h} = \lim_{h \to 0} \left(\frac{|h|^2}{h} \right) = \lim_{h \to 0} \bar{h} = 0.$$

For any $z_0 \neq 0$,

$$\frac{[f(z_0 + h) - f(z_0)]}{h} = \frac{[|z_0 + h|^2 - |z_0|^2]}{h}$$

$$= \frac{(h\bar{z}_0 + \bar{h}z_0 + |h|^2)}{h}$$

$$= \bar{z}_0 + \left(\frac{\bar{h}}{h} \right) z_0 + \bar{h}.$$

If we take h to be real and let $h \to 0$, then

$$\lim_{h \to 0} \frac{[f(z_0 + h) - f(z_0)]}{h} = \bar{z}_0 + z_0.$$

But if $h = ir$ for real r, and $r \to 0$, then

$$\lim_{h \to 0} \frac{[f(z_0 + h) - f(z_0)]}{h} = \bar{z}_0 - z_0.$$

If $z_0 \neq 0$, then

$$\bar{z}_0 + z_0 \neq \bar{z}_0 - z_0,$$

and $f(z)$ is not differentiable at z_0.

This example exhibits a function with a derivative at an isolated point. There are several reasons for suggesting that we should not be too impressed with the existence of a derivative at a single point and that we should restrict our attention to functions having a derivative at every point of a domain. The first reason comes from the properties of real power series in their regions of convergence. Recall that if such a power series in powers of $x - x_0$ has a radius of convergence $r > 0$, then the function represented by the series has a derivative at every point of the region of convergence $|x - x_0| < r$. However, it is usually difficult to deal in a practical way with a function known only by a power series representation with a positive radius of convergence. By turning our attention to functions defined and differentiable in a domain, rather than differentiable merely at isolated points, we shall eventually find that we are talking about exactly those functions which can be represented by power series with a positive radius of convergence. What we learn in the process will, among other things, help us work with functions defined by power series.

A second reason is that, in trying to exhibit conditions on a function $f(z)$ sufficient to ensure that $f(z)$ is differentiable at a point z_0, we shall see that we want $f(z)$ to behave nicely at every point in a neighborhood of z_0. The details will appear in Theorems 1.4 and 1.5. For these reasons we make the following definition.

Definition   A function $f(z)$ defined in a domain D containing the point z_0 is said to be *analytic* at z_0 if for some number $r > 0$, such that the open disk $\{z : |z - z_0| < r\}$ is contained in D, $f(z)$ is differentiable at each point of this disk.

Note that, although the definition of "analytic at z_0" suggests this to be a property for $f(z)$ at a point, this is really a property for $f(z)$ in a neighborhood of a point. In Example 1.5 we have a function which is differentiable at $z_0 = 0$ but not analytic at $z_0 = 0$.

Let $f(z)$ be defined in a domain D, where $z = x + iy$, $u(x,y) = \operatorname{Re} f(z)$, $v(x,y) = \operatorname{Im} f(z)$. We are looking for conditions on $u(x,y)$ and $v(x,y)$ which will guarantee that $f(z)$ has a derivative at every point of D. We first see what conditions differentiability imposes on $u(x,y)$ and $v(x,y)$.

THEOREM 1.4 Hypotheses: D is a domain containing $z_0 = x_0 + iy_0$ and $f(z)$ is a function defined and continuous in D such that $f(z)$ is differentiable at z_0.

Conclusions:
C1 $u(x,y) = \operatorname{Re} f(z)$ and $v(x,y) = \operatorname{Im} f(z)$ have first-order partial derivatives at (x_0,y_0).
C2   The partial derivatives of u and v at (x_0,y_0) satisfy the equations

$$u_x = v_y \quad \text{and} \quad u_y = -v_x. \tag{1.1}$$

Proof: Since $f'(z_0)$ exists, we write

$$f'(z_0) = \lim_{h \to 0} \frac{[f(z_0 + h) - f(z_0)]}{h}.$$

We take h to be real first, so that

$$f'(z_0) = \lim_{h \to 0} \left\{ \frac{[u(x_0 + h,y_0) - u(x_0,y_0)]}{h} + \frac{i[v(x_0 + h,y_0) - v(x_0,y_0)]}{h} \right\}$$

$$= \lim_{h \to 0} \left\{ \frac{[u(x_0 + h,y_0) - u(x_0,y_0)]}{h} \right\}$$

$$+ i \lim_{h \to 0} \left\{ \frac{[v(x_0 + h,y_0) - v(x_0,y_0)]}{h} \right\}.$$

Each limit here must exist (by Theorem 1.1) and $f'(z_0) = u_x(x_0,y_0) + iv_x(x_0,y_0)$.

If we now take $h = ir$, where r is real, a similar argument will show that $f'(z_0) = v_y(x_0,y_0) - iu_y(x_0,y_0)$. Equating the real and imaginary parts in the resulting expressions for $f'(z_0)$, we obtain (1.1).

The equations in (1.1) are known as the *Cauchy-Riemann equations;* they give necessary conditions for $f(z)$ to have a derivative at z_0. However, they are not sufficient to guarantee that $f(z)$ have a derivative at z_0 — that is, we can construct a function $f(z)$, defined in a domain containing a point z_0 at which $\operatorname{Re} f(z)$ and $\operatorname{Im} f(z)$ satisfy the Cauchy-

Riemann equations, which has no derivative at z_0. (See problem 1.20.)

Theorem 1.5 offers sufficient conditions for $f(z)$ to have a derivative at a point z_0. They are not the weakest possible such conditions, but they are easy to verify in practice and general enough for our purposes. The proof is not exciting, but it does review some worthwhile ideas from calculus.

THEOREM 1.5   Hypotheses: $f(z)$ is defined in a domain D containing the point $z_0 = x_0 + iy_0$.
H1 $u(x,y) = \operatorname{Re} f(z)$ and $v(x,y) = \operatorname{Im} f(z)$ have continuous first partial derivatives in a neighborhood $|z - z_0| < r$ about z_0 in D.
H2 $u(x,y)$ and $v(x,y)$ satisfy the Cauchy-Riemann equations at (x_0,y_0).

Conclusion: $f(z)$ is differentiable at z_0.

Proof: If $h = a + ib$, where $0 < |h| < r$,

$$\frac{[f(z_0 + h) - f(z_0)]}{h}$$

$$= \frac{[u(x_0 + a, y_0 + b) - u(x_0,y_0)] + i[v(x_0 + a, y_0 + b) - v(x_0,y_0)]}{a + ib}$$

$$= \left\{ \frac{u(x_0 + a, y_0 + b) - u(x_0, y_0 + b) + u(x_0, y_0 + b) - u(x_0,y_0)}{a + ib} \right\}$$

$$+ i \left\{ \frac{v(x_0 + a, y_0 + b) - v(x_0, y_0 + b) + v(x_0, y_0 + b) - v(x_0,y_0)}{a + ib} \right\}$$

$$= \left(\frac{a}{a + ib} \right) \left\{ \left[\frac{u(x_0 + a, y_0 + b) - u(x_0, y_0 + b)}{a} \right] \right.$$

$$+ i \left[\frac{v(x_0 + a, y_0 + b) - v(x_0, y_0 + b)}{a} \right] \right\}$$

$$+ \left(\frac{b}{a + ib} \right) \left\{ \left[\frac{u(x_0, y_0 + b) - u(x_0, y_0)}{b} \right] \right.$$

$$+ i \left[\frac{v(x_0, y_0 + b) - v(x_0, y_0)}{b} \right] \right\}.$$

If either a or b is zero, the above rearrangements are easier.

Since each of $u(x,y)$ and $v(x,y)$ has continuous first derivatives in $|z - z_0| < r$, we can use the mean-value theorem for derivatives to find real numbers t_1, t_2, t_3, t_4, depending on h, such that $0 < t_j < 1$, $j = 1, 2, 3, 4$, and

$$\frac{[f(z_0 + h) - f(z_0)]}{h}$$

$$= \left(\frac{a}{a + ib}\right)[u_x(x_0 + t_1 a, y_0 + b) + iv_x(x_0 + t_2 a, y_0 + b)]$$

$$+ \left(\frac{b}{a + ib}\right)[u_y(x_0, y_0 + t_3 b) + iv_y(x_0, y_0 + t_4 b)].$$

If we let

$$E_1(h) = u_x(x_0 + t_1 a, y_0 + b) - u_x(x_0, y_0)$$
$$E_2(h) = v_x(x_0 + t_2 a, y_0 + b) - v_x(x_0, y_0)$$
$$E_3(h) = u_y(x_0, y_0 + t_3 b) - u_y(x_0, y_0)$$
$$E_4(h) = v_y(x_0, y_0 + t_4 b) - v_y(x_0, y_0),$$

then $\lim_{h \to 0} E_j(h) = 0$ for $j = 1, 2, 3, 4$, because $\lim_{h \to 0} a = \lim_{h \to 0} b = 0$.

We rewrite

$$\frac{[f(z_0 + h) - f(z_0)]}{h}$$

$$= \left(\frac{a}{a + ib}\right)[u_x(x_0, y_0) + E_1(h) + iv_x(x_0, y_0) + i\, E_2(h)]$$

$$+ \left(\frac{b}{a + ib}\right)[u_y(x_0, y_0) + E_3(h) + iv_y(x_0, y_0) + i\, E_4(h)]$$

$$= \left(\frac{a}{a + ib}\right)[u_x(x_0, y_0) + E_1(h) + iv_x(x_0, y_0) + i\, E_2(h)]$$

$$+ \left(\frac{b}{a + ib}\right)[-v_x(x_0, y_0) + E_3(h) + iu_x(x_0, y_0) + i\, E_4(h)],$$

having used the hypothesis H2.

With yet another rewriting we obtain

$$\frac{[f(z_0 + h) - f(z_0)]}{h}$$

$$= \left(\frac{1}{a + ib}\right)\Big\{(a + ib)u_x(x_0, y_0) + (ia - b)v_x(x_0, y_0)$$

$$+ a[E_1(h) + i\, E_2(h)] + b[E_3(h) + i\, E_4(h)]\Big\}$$

$$= u_x(x_0, y_0) + iv_x(x_0, y_0)$$

$$+ \left(\frac{a}{a + ib}\right)[E_1(h) + i\, E_2(h)] + \left(\frac{b}{a + ib}\right)[E_3(h) + i\, E_4(h)].$$

Since $\lim_{h\to 0} E_j(h) = 0$, $j = 1, 2, 3, 4$, given any number $\epsilon > 0$ we can find a number $\delta(\epsilon,z_0) > 0$ such that whenever $|h| < \delta(\epsilon,z_0)$, we have $|E_j(h)| < \epsilon/4$. Then for $|h| < \delta(\epsilon,z_0)$,

$$\left| \frac{[f(z_0 + h) - f(z_0)]}{h} - [u_x(x_0,y_0) + iv_x(x_0,y_0)] \right|$$

$$\leq \left(\frac{|a|}{|a + ib|} \right)[|E_1(h)| + |E_2(h)|] + \left(\frac{|b|}{|a + ib|} \right)[|E_3(h)| + |E_4(h)|]$$

$$\leq |E_1(h)| + |E_2(h)| + |E_3(h)| + |E_4(h)| < \epsilon.$$

This means that $f'(z_0)$ exists and equals

$$u_x(x_0,y_0) + iv_x(x_0,y_0).$$

Example 1.7   Let $f(z) = e^x \cos y + ie^x \sin y$ for all z. Then $\mathrm{Re}\, f(z) = e^x \cos y = u(x,y)$, and $\mathrm{Im}\, f(z) = e^x \sin y = v(x,y)$ have continuous first partial derivatives satisfying the Cauchy-Riemann equations at any point $z_0 = x_0 + iy_0$, and $f'(z_0) = u_x(x_0,y_0) + iv_x(x_0,y_0) = f(z_0)$. (Note that when $z = x + i0, f(z) = f(x) = e^x$. This, together with the fact that $f'(z) = f(z)$, will surely influence us in defining an exponential function, e^z, in Chapter 2.)

We shall be primarily concerned with functions which are analytic at all points of a domain, or at all but finitely many points of a domain. For this reason we want to rephrase Theorems 1.4 and 1.5 slightly.

THEOREM 1.4′   Hypothesis: $f(z)$ is analytic in a domain D.

Conclusion: $u(x,y) = \mathrm{Re}\, f(z)$ and $v(x,y) = \mathrm{Im}\, f(z)$ have first partial derivatives which satisfy the Cauchy-Riemann equations at each point of D.

THEOREM 1.5′   Hypotheses: $f(z)$ is defined in a domain D.
H1 $u(x,y) = \mathrm{Re}\, f(z)$ and $v(x,y) = \mathrm{Im}\, f(z)$ have continuous first partial derivatives at every point of D.
H2 $u_x(x,y) = v_y(x,y)$ and $u_y(x,y) = -v_x(x,y)$ at every point of D.

Conclusion: $f(z)$ is analytic in D.

In Theorem 1.4′ we should like to conclude that the first derivatives of $u(x,y)$ and $v(x,y)$ are continuous at each point of D, since if this were the case, Theorems 1.4′ and 1.5′ would be converse to each other. This is actually the case, although we shall have finished Chapter 3 by the time we are able to prove it.

There is one more theorem with a familiar flavor to it which we should examine at this time. Its purpose is to give us a useful "chain rule" for derivatives and convince us that (speaking imprecisely) "an analytic function of an analytic function is analytic."

THEOREM 1.6 Hypotheses:
H1 $g(z)$ is analytic in a domain D with range E.
H2 $f(w)$ is analytic in a domain containing E.

Conclusions:
C1 $F(z) = f[g(z)]$ is analytic in D.
C2 For each z_0 in D, $F'(z_0) = f'[g(z_0)]g'(z_0)$.

Proof: It is enough to show that C2 is true for an arbitrary point z_0 in D.

Since $f(w)$ is differentiable at $w_0 = g(z_0)$, it follows that

$$\lim_{k \to 0} [f(w_0 + k) - f(w_0)]/k = f'(w_0),$$

and so we can write $f(w_0 + k) - f(w_0) = f'(w_0)k + E_1(k)k$, where $\lim_{k \to 0} E_1(k) = 0$.

If h is any complex number such that $z_0 + h$ is also in D, let $k = g(z_0 + h) - g(z_0)$. Since $g(z)$ is differentiable at z_0,

$$\lim_{h \to 0} [g(z_0 + h) - g(z_0)]/h = g'(z_0),$$

and we write

$$k = g(z_0 + h) - g(z_0) = g'(z_0)h + E_2(h)h$$

where $\lim_{h \to 0} E_2(h) = 0$.

Also $g(z)$ is continuous at z_0, so $\lim_{h \to 0} k = 0$. This means that $\lim_{h \to 0} E_1(k) = \lim_{k \to 0} E_1(k) = 0$.

Now

$$\frac{[F(z_0 + h) - F(z_0)]}{h}$$

$$= \frac{\{f[g(z_0 + h)] - f[g(z_0)]\}}{h}$$

$$= \frac{[f(w_0 + k) - f(w_0)]}{h}$$

$$= \frac{[f'(w_0)k + E_1(k)k]}{h}$$

$$= \frac{\{f'(w_0)[g'(z_0)h + E_2(h)h] + E_1(k)[g'(z_0)h + E_2(h)h]\}}{h}$$

$$= f'(w_0)g'(z_0) + f'(w_0)E_2(h) + g'(z_0)E_1(k) + E_1(k)E_2(h).$$

If we now take limits as $h \to 0$, we see that

$$F'(z_0) = f'[g(z_0)]g'(z_0).$$

PROBLEMS

1.20 Show in two ways that $f(z) = f(x + iy) = 3x + i4y$ is not differentiable at $z = 0$.

1.21 Verify that $f(z) = (|xy|)^{1/2}$ satisfies the Cauchy-Riemann equations at $z = 0$ but does not have a derivative at $z = 0$.

1.22 Suppose $f(z) = \sin x \cosh y + i \cos x \sinh y$. Find $f'(z_0)$ at any point $z_0 = x_0 + iy_0$. If $z_0 = x_0 + i0$, what is $f(z_0)$ and what is $f'(z_0)$?

1.23 If n is a positive integer and $f(z) = z^n$, verify that $f'(z_0) = n z_0^{n-1}$ for any z_0.

1.24 Suppose $f(z) = u(x,y) + i v(x,y)$, where u and v in polar coordinates have continuous first partial derivatives with respect to r and θ at some point $z_0 = r(\cos \theta + i \sin \theta)$, $z_0 \neq 0$. Show that if $u_r = (1/r)v_\theta$ and $v_r = -(1/r)u_\theta$, then $f(z)$ is analytic at z_0, and $f'(z_0) = (\cos \theta - i \sin \theta)(u_r + iv_r)$.

1.25 If $z = r(\cos \theta + i \sin \theta)$, $r > 0$, $0 < \theta < 2\pi$, and $f(z) = r^{1/2}[\cos (\theta/2) + i \sin (\theta/2)]$, verify that the equations of the preceding problem are satisfied and that

$$f'(z) = \left(\frac{1}{2}r^{1/2}\right)[\cos (\theta/2) - i \sin (\theta/2)].$$

1.26 We have seen that if $f(z) = u(x,y) + iv(x,y)$ is differentiable at $z_0 = x_0 + iy_0$, then $f'(z_0) = u_x(x_0,y_0) + iv_x(x_0,y_0) = v_y(x_0,y_0) - iu_y(x_0,y_0)$. Let $\mathbf{a}$ be a unit vector at z_0 making an angle α with the positive real axis. Show that

$$f'(z_0) = [D_\mathbf{a} u(x_0,y_0) + iD_\mathbf{a}v(x_0,y_0)](\cos \alpha - i \sin \alpha)$$

where $D_a u(x_0,y_0)$, $D_a v(x_0,y_0)$ are the directional derivatives of u and v at (x_0,y_0) in the direction of **a**. Notice the equations for $f'(z_0)$ obtained by choosing $\alpha = 0$ and $\alpha = \pi/2$.

1.27 If $f(z)$ is defined for $|z - z_0| < r$ and $|f(z)|$ is constant there, then show that if $f(z)$ is analytic there it must be constant.

1.28 Show that if $f(z)$ is defined, nonconstant, and real-valued for $|z - z_0| < r$, then $f(z)$ cannot be analytic at z_0. (However, as in Example 1.5, $f(z)$ might have a derivative at z_0.)

1.29 Does the function $f(z) = \bar{z}$ have a derivative anywhere? Is it analytic anywhere?

1.30 If $f(z)$ is analytic in a domain D and $f'(z) = 0$ at every point of D, show that $f(z)$ is identically constant in D.

Section 1.4   Harmonic Functions

If D is a domain in R^2 and if g is a real-valued function defined on D having second partial derivatives at each point of D, we call g *harmonic in D* if

$$\nabla^2 g(x,y) = g_{xx}(x,y) + g_{yy}(x,y) = 0 \qquad (1.2)$$

at every point (x,y) of D. Here $\nabla^2 g$ is the *Laplacian of g*, and Equation (1.2) is known as *Laplace's equation* in two dimensions. Such functions arise quite naturally when we deal with analytic functions of a complex variable.

If $f(z) = u(x,y) + iv(x,y)$ is analytic in a domain D of the complex plane, we know that u and v have first partial derivatives at each point of D, but we do not yet know that these derivatives are themselves continuous or possess first partial derivatives. For the present let us assume that u and v have continuous second partial derivatives at each point where $f(z) = u(x,y) + iv(x,y)$ is analytic. Our work in Chapter 3 will show us that u and v will have these properties.

THEOREM 1.7 Hypotheses:
H1 $f(z)$ is analytic in a domain D.
H2 $f(z) = u(x,y) + iv(x,y)$, where u and v have continuous second partial derivatives at each point of D.

Conclusion: $u(x,y)$ and $v(x,y)$ are harmonic in D.

Proof: We make a proof for $u(x,y)$ and let the reader supply the details for $v(x,y)$.

At any point $z_0 = x_0 + iy_0$ of D, H1 implies that

$$u_x(x_0,y_0) = v_y(x_0,y_0) \quad \text{and} \quad u_y(x_0,y_0) = -v_x(x_0,y_0).$$

Thus

$$u_{yy}(x_0,y_0) = -v_{xy}(x_0,y_0) = -v_{yx}(x_0,y_0) = -u_{xx}(x_0,y_0).$$

(Why are these steps valid?) Then

$$\nabla^2 u(x_0,y_0) = u_{xx}(x_0,y_0) + u_{yy}(x_0,y_0) = 0.$$

The reader might take time now to verify that the real and imaginary parts of the functions defined in Example 1.7 and problem 1.22 are indeed harmonic.

For harmonic functions in domains of R^2 our work with analytic functions can tell us something further: Harmonic functions come in pairs, the "binding agent" in each pair being an analytic function. We state the situation in Theorem 1.8, but without giving a proof.

THEOREM 1.8   Hypothesis: $u(x,y)$ is harmonic in a simply connected domain D.

Conclusion: There exists a function $f(z)$, analytic in D, such that $u(x,y) = \operatorname{Re} f(z)$ at each point of D.

Comments:

1.   The imaginary part of $f(z)$ here is also harmonic in D; we call it a *harmonic conjugate* to $\operatorname{Re} f(z)$. Thus Theorem 1.8 might be rephrased to say: If $u(x,y)$ is harmonic in D, there exists a conjugate harmonic function $v(x,y)$ to $u(x,y)$ such that $f(z) = u(x,y) + iv(x,y)$ is analytic in D.

2.   Theorem 1.8 can fail to be true if D is not simply connected. (See problem 2.26.)

We can illustrate a procedure for finding a harmonic conjugate to a given harmonic function with an example. The justification of the steps of this procedure would go far toward actually proving Theorem 1.8. Perhaps the reader will recognize the procedure from discussions of Green's Theorem in the plane or methods for solving first-order, exact, linear, ordinary differential equations.

Example 1.8 $u(x,y) = y^3 - 3x^2y$ is harmonic for all $z = x + iy$. For any z and any simply-connected domain containing z, we know there exists a function $f(z) = u(x,y) + iv(x,y)$ analytic in this domain satisfying $u_x(x,y) = v_y(x,y)$ and $u_y(x,y) = -v_x(x,y)$.

Now if $u_x = -6xy = v_y$, then $v(x,y) = -3xy^2 + h(x)$, where $h(x)$ is some real-valued function of x. Also

$$v_x = -3y^2 + h'(x) = -u_y,$$

so

$$u_y = 3y^2 - h'(x) = 3y^2 - 3x^2.$$

Then $h'(x) = 3x^2$, $h(x) = x^3 + k$, where k is an arbitrary constant. We set $v(x,y) = -3xy^2 + x^3 + k$; $v(x,y)$ is harmonic, and it is easily verified that $f(z) = u(x,y) + iv(x,y) = (y^3 - 3x^2y) + i(x^3 - 3xy^2) + ik = z^3 + ik$ is analytic.

PROBLEMS

1.31 Verify that $u(x,y)$ is harmonic and find a harmonic conjugate to $u(x,y)$.
   (a) $u(x,y) = y/(x^2 + y^2)$, $(x,y) \neq (0,0)$.
   (b) $u(x,y) = \sin x \cosh y$
   (c) $u(x,y) = x^2 - y^2$.

1.32 Suppose $u(x,y)$ is harmonic in a domain D and $v(x,y)$ is a harmonic conjugate to $u(x,y)$ in D. For real constants c and k, the equations $u(x,y) = c$, $v(x,y) = k$ determine families of what are called *level curves* for u and v.

   Suppose for $c = c_0$, $k = k_0$ that the level curves $u(x,y) = c_0$, $v(x,y) = k_0$ intersect at a point $z_0 = x_0 + iy_0$ of D. Prove that if the function $f(z) = u(x,y) + iv(x,y)$ is analytic at z_0, with $f'(z_0) \neq 0$, then the level curves $u(x,y) = c_0$, $v(x,y) = k_0$ are orthogonal at z_0.

1.33 If $u(x,y) = x^2 - y^2$, $v(x,y) = 2xy$, sketch a few representative level curves for each function. Note that the curves defined by $u(x,y) = 0$ and $v(x,y) = 0$ intersect at $(0,0)$ but are not orthogonal. Why isn't this a contradiction to the preceding problem?

1.34 Suppose D is a domain which does not contain $(0,0)$ and $u(x,y)$ is a real-valued function with continuous second partial derivatives in D.

Show that under the change of variables $\{x = r\cos\theta, y = r\sin\theta\}$, $\nabla^2 u = u_{xx} + u_{yy}$ becomes

$$\nabla^2 u = u_{rr} + \left(\frac{1}{r}\right)u_r + \left(\frac{1}{r^2}\right)u_{\theta\theta}.$$

2

Elementary Functions

In this chapter we want to arrive at natural definitions for complex-valued functions of a complex variable corresponding to the exponential, logarithmic, trigonometric, and hyperbolic functions dealt with so frequently in elementary calculus. Of course, if we are going to define what we mean by sin z for a complex number z, the definition we choose should be consistent with what we already understand sin z to mean when z is a real number. If we try to define sin z for complex z so as to preserve the familiar properties of sin z for real z, we find that our definition is virtually determined by the demands of consistency with the real case. It is in this sense that our definitions will be natural.

Section 2.1 The Exponential Function

If we want a function $f(z)$ of a complex variable z to have the property that $f(z) = e^z$ when z is real, we shall also want $f'(z) = f(z)$, and

$f(z_1 + z_2) = f(z_1)f(z_2)$. In Example 1.6 we exhibited a function $f(z)$ with the first two of these properties. To determine how such a function might have been expected to serve as a definition for e^z for complex z, we start with the Maclaurin's series for e^u, where u is a real variable, and formally replace u by iy, where y is real:

$$e^{iy} = \sum_{j=0}^{\infty} \frac{(iy)^j}{j!} = \sum_{k=0}^{\infty} \frac{(iy)^{2k}}{(2k)!} + \sum_{k=0}^{\infty} \frac{i^{2k+1}y^{2k+1}}{(2k+1)!}$$

$$= \sum_{k=0}^{\infty} \frac{(-1)^k y^{2k}}{(2k)!} + i \sum_{k=0}^{\infty} \frac{(-1)^k y^{2k+1}}{(2k+1)!},$$

since $i^{2k} = (-1)^k$ and $i^{2k+1} = i^{2k}i = i(-1)^k$ for each nonnegative integer k. The last two series above are the formal Maclaurin's series for $\cos y$ and $\sin y$, respectively, so we write $e^{iy} = \cos y + i \sin y$. Now for $z = x + iy$, we should expect that $e^z = e^x e^{iy} = e^x(\cos y + i \sin y)$, and this is the function $f(z)$ of Example 1.6.

In the paragraph above we have been insensitive to the mathematical niceties and all problems of validity, but at least we have some feeling that we should define the exponential function, e^z, for all $z = x + iy$ in the following way.

Definition $e^z = e^{x+iy} = e^x(\cos y + i \sin y) = \exp z$.

By Example 1.6, e^z is analytic for all z, with $(e^z)' = e^z$; we also see that $|e^z| = e^x$.

Let us summarize a number of properties for e^z. The proofs are left to the reader.

1. For any z_1, z_2, $\exp(z_1 + z_2) = (\exp z_1)(\exp z_2)$.
2. For any z, $e^z \neq 0$.
3. $1/e^z = e^{-z}$.
4. For any z, $\exp(z + 2\pi i) = \exp z$.
5. For any z and any real rational number r, $(e^z)^r = e^{rz}$.

Further properties for e^z appear in the problems.

If $z = x + iy$ is written in polar form as $z = r(\cos \theta + i \sin \theta)$ for $r = (x^2 + y^2)^{1/2} \neq 0$, we may now rewrite this polar form as $z = r \exp(i\theta)$. Consequently, $\bar{z} = r \exp(-i\theta)$; and if $z_1 = r_1 \exp(i\theta_1)$, $z_2 = r_2 \exp(i\theta_2)$, $r_2 \neq 0$, we see that

$$z_1 z_2 = r_1 r_2 \exp[i(\theta_1 + \theta_2)]$$

$$\frac{z_1}{z_2} = \left(\frac{r_1}{r_2}\right) \exp[i(\theta_1 - \theta_2)].$$

PROBLEMS

2.1   Verify properties 1–5 above for e^z.

2.2   Show that $\exp \bar{z} = \overline{\exp z}$.

2.3   For what values of z will $\exp(i\bar{z}) = \overline{\exp(iz)}$?

2.4   Give an example to show that $(\exp z)^w$ and $\exp(zw)$ need not be equal.

2.5   Find all values of z for which
   (a) $e^z = 5i$
   (b) $e^z = 1 + i$
   (c) $|\exp(-iz)| > 1$.

2.6   If $g(z)$ is analytic at z_0 and $F(z) = \exp[g(z)]$, verify that $F(z)$ is analytic at z_0 and that $F'(z_0) = F(z_0)g'(z_0)$.

2.7   We defined $f(z) = e^z = e^x(\cos y + i \sin y)$ because $f(z)$ satisfied the conditions: (a) $f(z)$ is analytic for all z; (b) $f'(z) = f(z)$; (c) $f(z) = e^z$ for z real. That is, we implicitly assumed that there is only one function $f(z)$ satisfying (a), (b), (c), and that it must be $f(z) = e^x(\cos y + i \sin y)$.

   Prove that if $f(z)$ satisfies (a), (b), (c), then necessarily $f(z) = e^x(\cos y + i \sin y)$.

2.8   Suppose that $f(z)$ satisfies (a) and (c) of the preceding problem and also has the property: $f(z_1 + z_2) = f(z_1)f(z_2)$. Prove that $f(z)$ must then have the property that $f'(z) = f(z)$ for all z.

Section 2.2   The Trigonometric and Hyperbolic Functions

For real y, let us solve this pair of equations for $\cos y$ and $\sin y$:

$$e^{iy} = \cos y + i \sin y, \qquad e^{-iy} = \cos y - i \sin y.$$

We find that $\cos y = (\tfrac{1}{2})(e^{iy} + e^{-iy})$ and $\sin y = (1/2i)(e^{iy} - e^{-iy})$. This leads us to make the following definition.

Definition   For each complex number z

$$\sin z = \frac{e^{iz} - e^{-iz}}{2i} \qquad \cos z = \frac{e^{iz} + e^{-iz}}{2}$$

Further, whenever the denominators involved are not zero, define

$$\tan z = \frac{\sin z}{\cos z}, \quad \csc z = \frac{1}{\sin z}, \quad \sec z = \frac{1}{\cos z}, \quad \cot z = \frac{1}{\tan z}$$

These definitions yield the usual real-valued trigonometric functions when z is real, and they show how important it is for us to know exactly when $\sin z = 0$ and $\cos z = 0$. For example, we know $\sin z = 0$ when $z = n\pi$ for any integer n, but perhaps there are other numbers in the complex plane where $\sin z = 0$. (We shall see below that this is not the case.)

It is easy to see, with repeated use of Theorem 1.6, that whenever these six functions are defined they are analytic, and

$$(\sin z)' = \cos z; \ (\cos z)' = -\sin z; \ (\tan z)' = \sec^2 z;$$

$$(\csc z)' = -\csc z \cot z; \ (\sec z)' = \sec z \tan z;$$

$$(\cot z)' = -\csc^2 z.$$

For $z = x + iy$, $e^{iz} = e^{-y}e^{ix}$ and $e^{-iz} = e^{y}e^{-ix}$; then

$$\sin z = \frac{e^{-y}e^{ix} - e^{y}e^{-ix}}{2i}$$

$$= \frac{(e^{-y} - e^{y})\cos x + i(e^{-y} + e^{y})\sin x}{2i}$$

Hence

$$\sin(x + iy) = \sin x \cosh y + i \cos x \sinh y.$$

Similarly we can show that

$$\cos(x + iy) = \cos x \cosh y - i \sin x \sinh y.$$

From these equations it is easy to derive the following relations:

$$\sin(iy) = i \sinh y, \ \cos(iy) = \cosh y \text{ for real } y;$$

$$\sin \bar{z} = \overline{\sin z}, \ \cos \bar{z} = \overline{\cos z}.$$

Furthermore, the familiar identities from trigonometry will hold. We list a few of these identities for illustration:

$$\sin^2 z + \cos^2 z = 1$$

$$\sin(-z) = -\sin z \qquad \cos(-z) = \cos z$$

$$\sin(z_1 + z_2) = \sin z_1 \cos z_2 + \cos z_1 \sin z_2$$

$$\cos(z_1 + z_2) = \cos z_1 \cos z_2 - \sin z_1 \sin z_2.$$

If $\sin z = 0$ for $z = x + iy$, then both $\sin x \cosh y = 0$ and $\cos x \sinh y = 0$. From the first equation we see that $x = n\pi$ for some integer n, since $\cosh y$ is never zero. But $\cos x \neq 0$ for $x = n\pi$, so for the second equation to hold it must be that $\sinh y = 0$, which occurs only when $y = 0$. Thus $\sin z = 0$ only when $z = n\pi$ for some integer n; that is, the function $\sin z$ has only real zeros. The reader should show for himself that $\cos z = 0$ only when $z = k\pi/2$, for k any odd integer.

We have extended the domains of $\sin z$ and $\cos z$ and the other trigonometric functions from the real axis to the complex plane. We see that for $\sin z$ and $\cos z$ our extended definitions have not added any additional points where these functions vanish, and for all six functions we have preserved the familiar behavior of the real-valued trigonometric functions.

However, one pleasant feature of $\sin z$ and $\cos z$ for real z has been sacrificed: If z is not real, we can no longer assume that $|\sin z| \leq 1$ and $|\cos z| \leq 1$. For

$$\begin{aligned}
|\sin z|^2 &= \sin^2 x \cosh^2 y + \cos^2 x \sinh^2 y \\
&= \sin^2 x(1 + \sinh^2 y) + \cos^2 x \sinh^2 y \\
&= \sin^2 x + (\sin^2 x + \cos^2 x)\sinh^2 y \\
&= \sin^2 x + \sinh^2 y.
\end{aligned}$$

Similarly

$$|\cos z|^2 = \cos^2 x + \sinh^2 y.$$

Since $\sinh y$ is not bounded above or below, we can make $|\sin z|$ or $|\cos z|$ arbitrarily large merely by taking $\operatorname{Im} z = y$ sufficiently far from zero. One implication of this fact is that we can now expect equations like $\sin z = 2$ or $\cos z = 57$ to have solutions. (We can be sure, however, that none of the solutions is real.)

Following the pattern above for trigonometric functions, we are led to define hyperbolic functions of a complex variable z in the following way.

Definition   For each complex number z,

$$\sinh z = \frac{(e^z - e^{-z})}{2}, \qquad \cosh z = \frac{(e^z + e^{-z})}{2}.$$

As long as the denominators involved are not zero, we also define

$$\tanh z = \frac{\sinh z}{\cosh z} \qquad \operatorname{csch} z = \frac{1}{\sinh z}$$

$$\operatorname{sech} z = \frac{1}{\cosh z} \qquad \coth z = \frac{1}{\tanh z}$$

Of course when z is real, these equations reduce to the real-valued hyperbolic functions. Because of the properties of the exponential function, e^z, wherever these hyperbolic functions are defined they are analytic. As with the trigonometric functions, it is important to know exactly where these functions fail to be defined — that is, where $\sinh z = 0$ or $\cosh z = 0$. We expect $\sinh z = 0$ for $z = 0$, and we expect $\cosh z = 0$ to have no real solutions; however, either equation may have additional solutions off the real axis.

For $z = x + iy$,

$$\sinh z = \sinh(x + iy)$$

$$= \frac{[\exp(x + iy) - \exp(-x - iy)]}{2}$$

$$= \frac{[(e^x - e^{-x})\cos y + i(e^x + e^{-x})\sin y]}{2},$$

so that

$$\sinh(x + iy) = \sinh x \cos y + i \cosh x \sin y.$$

Similarly,

$$\cosh(x + iy) = \cosh x \cos y - i \sinh x \sin y.$$

The reader should now solve for himself the equations $\sinh z = 0$ and $\cosh z = 0$. From the equations above — and with varying amounts of manipulation — we can derive additional identities for the hyperbolic functions. The following are representative:

$$\sinh(iy) = i \sin y, \qquad \cosh(iy) = \cos y$$

$$|\sinh z|^2 = \sinh^2 x + \sin^2 y$$

$$|\cosh z|^2 = \sinh^2 x + \cos^2 y$$

$$\sinh(z_1 + z_2) = \sinh z_1 \cosh z_2 + \cosh z_1 \sinh z_2.$$

The derivatives for the hyperbolic functions, which exist wherever these functions are defined, are easily verified to be:

$$(\sinh z)' = \cosh z, \qquad (\cosh z)' = \sinh z,$$

$$(\tanh z)' = \operatorname{sech}^2 z, \qquad (\coth z)' = -\operatorname{csch}^2 z,$$

$$(\operatorname{csch} z)' = -\operatorname{csch} z \coth z, \qquad (\operatorname{sech} z)' = -\operatorname{sech} z \tanh z.$$

PROBLEMS

2.9   Show that $\cos z = 0$ if and only if $z = n\pi/2$, where n is any odd integer.

2.10 Find all the solutions to the equations:
(a) $\cos z = 4$
(b) $\sin z = 2i$
(c) $2ie^{iz} = \sin z$

2.11 If $z = x + iy$, verify that $|\cos z|^2 = (\cosh 2y + \cos 2x)/2$.

2.12 Find all the solutions to the equations:
(a) $\sinh z = 0$
(b) $\cosh z = 0$
(c) $\sinh z = 2i$
(d) $\cosh z = 4$

2.13 Verify that $\sin(z + 2\pi) = \sin z$, $\cos(z + 2\pi) = \cos z$ for all z.

2.14 Verify that $\cos z = \sin(\pi/2 - z)$ for all z.

2.15 Verify that $\sinh(z + 2\pi i) = \sinh z$, $\cosh(z + 2\pi i) = \cosh z$ for all z.

2.16 Verify that $\cosh z = -i \sinh(\pi i/2 - z)$ for all z.

2.17 If x and y are real-valued, solve the equation $y = \sinh x$ for x and explain why the square root in your answer must be the positive square root.

Section 2.3   Inverse Functions and Multiple-Valued Functions

If, instead of trying to solve the equation $\sin z = 2i$, we try to solve the equation $\sin z = w$, for any given complex number w, we should like to know at least that there are some solutions to find. Then we should like to have a rather general solution of the form $z = f(w)$ which will give us solutions to the equation when we are given w. We have written the form of the solution as if it might be a function; but if there are many solutions for z to the equation $\sin z = w$, it might require many functions in order to solve the equation $\sin z = w$ in general. Since all of these "solution functions" would have the common property of solving a certain equation, we would like to find a way of lumping them together into one entity, which we shall come to call a multiple-valued function. The phrase "multiple-valued function" is convenient, but confusing, since such an entity is technically not a bonafide function.

We plan to be somewhat systematic in arriving at a number of equation-solving entities which suggest the notion of an inverse function of an elementary function. The inverse trigonometric functions will be among the last we consider, because the collections of functions they represent are quite difficult to manipulate.

The Logarithmic Functions

Since e^z is never zero, the equation $w = e^z$ will never have a solution for z corresponding to $w = 0$. For w different from zero, write $w = |w| \exp(i \arg w)$; here $\arg w$ may have any of an infinite number of values. Let θ be a particular value of $\arg w$, so that $w = |w| \exp(i\theta)$. If we take $z = \text{Log } |w| + i\theta$, where $\text{Log } |w|$ denotes the natural logarithm of the positive number $|w|$, then $\exp z = \exp(\text{Log } |w|)\exp(i\theta) = |w| \exp(i\theta) = w$, and this z is a solution to the equation $w = e^z$. For any choice of a value for $\arg w$ we get such a solution.

Definition   For any complex number z, $z \neq 0$, a *logarithm of* z will be any number $w = \text{Log } |z| + i \arg z$, where $\text{Log } |z|$ is the natural logarithm of $|z|$ and $\arg z$ is any value for the argument of z.

At this point we seem to have too many logarithms, and we certainly have nothing we can call a "logarithm function."

If z is a positive real number, of all the logarithms of z defined, there is just one which coincides with the natural logarithm of z: the logarithm of z corresponding to the choice $\arg z = 0$. For any z, $z \neq 0$, there is exactly one value of arg z such that $-\pi < \arg z \leq \pi$; we called this the principal argument of z in Chapter 1.

Definition   For each complex number z, $z \neq 0$, the *principal value of the logarithm of z* is defined as

$$\text{Log } z = \text{Log } |z| + i \text{ Arg } z.$$

The function $w = \text{Log } z$, defined for all $z \neq 0$, is called the *principal logarithm function*.

For each integer k we could define another logarithm function corresponding to the choice for arg z: $(2k - 1)\pi < \arg z \leq (2k + 1)\pi$. The collection of all these logarithm-like functions (and there are infinitely many) is commonly called the logarithm function, and each function in the collection is called a *branch* of the logarithm function.

Definition   The (multiple-valued) *logarithm function* is the infinite collection of branches

$$\log z = \text{Log } |z| + i \arg z, \qquad z \neq 0$$

$$(2k - 1)\pi < \arg z \leq (2k + 1)\pi, \qquad k = 0, \pm 1, \pm 2, \ldots$$

The branch corresponding to $k = 0$ is called the *principal branch of* log z, or the *principal logarithm function*, and is denoted by Log z.

Admittedly the language, especially the two uses of the word "function," is confusing. Each branch of the collection defining the logarithm function is a function defined for all $z \neq 0$.

We introduce the following conventions: the statement that $z \leq 0$ means that z is real and non-positive; $z < 0$ means that z is real and negative; $z \geq 0$ and $z > 0$ are defined similarly.

THEOREM 2.1   Each branch, log z, of the logarithm function has the following properties:
  (1)   log z is discontinuous for each $z \leq 0$;
  (2)   log z is analytic for all z except $z \leq 0$, and $(\log z)' = 1/z$;
  (3)   for all z, $z \neq 0$, any branch of the logarithm function differs from any other branch by an integral multiple of $2\pi i$.

Proof: Choose any integer k and keep it fixed; we look at the branch of the logarithm function corresponding to $(2k - 1)\pi < \arg z \leq (2k + 1)\pi$.

For any point $z_0 < 0$, as z approaches z_0 from the lower half plane, arg z approaches arg $z_0 - 2\pi$; as z approaches z_0 from the upper half plane, arg z approaches arg z_0. Thus, for z in the lower half plane,

$$\lim_{z \to z_0} \log z = \lim_{z \to z_0} \text{Log } |z| + i \lim_{z \to z_0} \text{arg } z$$

$$= \text{Log } |z_0| + i(\text{arg } z_0 - 2\pi) = \log z_0 - 2\pi i.$$

For z in the upper half plane, $\lim_{z \to z_0} \log z = \log z_0$, so $\log z$ must be discontinuous at z_0, and this proves (1).

To prove (2), let z be any non-zero point not lying on the negative real axis, and for any integer k write $z = r \exp(i\theta)$, where $r > 0$, $(2k - 1)\pi < \theta < (2k + 1)\pi$. Then $\log z = \text{Log } r + i\theta$, and we can see that $\log z$ satisfies the Cauchy-Riemann equations in polar form. (See problem 1.23.) This, and the continuity of $\text{Log } r$ and θ as functions of r and θ for $r > 0$ and $(2k - 1)\pi < \theta < (2k + 1)\pi$, implies that $\log z$ has a derivative at each such z, and

$$(\log z)' = (\cos \theta - i \sin \theta)\left[\left(\frac{1}{r}\right) + i0\right]$$

$$= \left(\frac{1}{r}\right)\exp(-i\theta) = \frac{1}{z}.$$

The conclusion (3) is obviously satisfied.

The line $\{z: z \leq 0\}$, where all the branches of the logarithm function fail to be analytic or even continuous, is called a *branch cut* for the branches of the logarithm.

As long as we take some care in dealing with the arguments of complex numbers, we can expect the usual properties of a logarithm function for real positive numbers to be valid in the complex plane. To illustrate, let $z_1 = \exp(i\pi/2)$, $z_2 = \exp(i3\pi/4)$. By using the principal branch of the logarithm, we see that

$$\text{Log } z_1 = \frac{i\pi}{2}, \qquad \text{Log } z_2 = \frac{i3\pi}{4},$$

$$\text{Log}(z_1 z_2) = \text{Log } \exp\left(\frac{i5\pi}{4}\right) = \text{Log } \exp\left(\frac{-i3\pi}{4}\right) = \frac{-i3\pi}{4},$$

so that

$$\text{Log}(z_1 z_2) = \text{Log } z_1 + \text{Log } z_2 - 2\pi i.$$

Suppose z_1 and z_2 are nonzero complex numbers. For any branch of the logarithm function we cannot expect to have $\log(z_1 z_2)$ equal $\log z_1 + \log z_2$ or to have $\log(z_1/z_2)$ equal $\log z_1 - \log z_2$ in general. However, we can see that the differences, $\log(z_1 z_2) - [\log z_1 + \log z_2]$, and $\log(z_1/z_2) - [\log z_1 - \log z_2]$, must either be zero or an integral multiple of $2\pi i$.

If we say, for two complex numbers A and B, that $A = B \pmod{2\pi i}$ if $A - B$ is an integral multiple of $2\pi i$, then in this sense the usual properties of logarithms are valid for each branch of the logarithm function:

$$\log(z_1 z_2) = \log z_1 + \log z_2 \qquad \text{(modulo } 2\pi i\text{);}$$

$$\log\left(\frac{z_1}{z_2}\right) = \log z_1 - \log z_2 \qquad \text{(modulo } 2\pi i\text{).}$$

Functions of the form z^a

In elementary calculus we defined x^a by the equation $x^a = \exp(a \operatorname{Log} x)$, where a is real, $x > 0$, and $\operatorname{Log} x$ denotes the natural logarithm of x. The following definition follows from this equation.

Definition If z is a nonzero complex number and a is any complex number, we define $z^a = \exp(a \log z)$. We call $z^a = \exp(a \operatorname{Log} z)$ the *principal value* of z^a. As z varies over the nonzero complex numbers, each branch of the logarithm function determines a branch for z^a.

Citing Theorems 1.6 and 2.1, as well as the properties of the exponential function, the reader can prove this statement.

THEOREM 2.2 Hypotheses: Let a and b be any complex numbers and let $\log z$ denote any particular branch of the logarithm function.

Conclusions:
C1   The corresponding branch of z^a is analytic where $\log z$ is analytic.
C2 $(z^a)' = az^{a-1}$, $z \neq 0$.
C3   for each nonzero z, $z^a z^b = z^{a+b}$.
C4   for any nonzero z_1, z_2, $(z_1 z_2)^a = z_1^a z_2^a \exp(2a\pi ki)$ for some integer k.
C5   for each nonzero z, $z^{-a} = 1/z^a$.

Of course z^a has a value corresponding to each possible value for $\log z$, but the periodicity of the exponential function indicates that distinct values of $\log z$ need not determine distinct values of z^a. We examine three cases where a is real.

Case I   a is an integer

Since for any z, $z \neq 0$, distinct branches of the logarithm function differ by an integral multiple of $2\pi i$, so do distinct branches of $a \log z$, and z^a has only one value given by $\exp(a \text{ Log } z)$. By defining $z^a = 0$ for $z = 0$, the reader can show z^a has a unique value for every z and is analytic for all z. (What has become of the discontinuities of $\text{Log } z$ on the negative real axis here?)

Case II   a is a rational number

Let $a = p/q$, where p is an integer, q is a positive integer, and p/q is reduced to lowest terms. For any value of $\arg z$, $z^a = \exp(a \log z) = \exp[(p/q)\log |z|]\exp[i(p/q)\arg z]$, and the first factor is independent of the value of $\arg z$. If $\alpha = \text{Arg } z$, then all values of $\arg z$ can be expressed as:

$$\arg z = \alpha + 2k\pi, \qquad k = 0, \pm 1, \pm 2, \ldots,$$

and

$$(p/q)\arg z = (p/q)\alpha + (p/q)2k\pi,$$

$$\exp[i(p/q)\arg z] = \exp[i(p/q)\alpha]\exp[i(p/q)2k\pi].$$

Now $\exp[i(p/q)2k\pi]$ takes on distinct values for $k = 0, 1, 2, \ldots, q - 1$, but for any other value of k yields one of the q values already produced. Thus $\exp[i(p/q)\arg z]$, and z^a, has only q distinct values. If we write $\log z = \text{Log } z + 2k\pi i$, $k = 0, 1, 2, \ldots, q - 1$,

$$z^a = z^{p/q} = \exp\left[\left(\frac{p}{q}\right)\text{Log } z\right]\exp\left[i\left(\frac{p}{q}\right)2k\pi\right]$$

give the q distinct values of z^a.

Example 2.1   Let $f(z) = z^{1/2}$. Here $p = 1$, $q = 2$, and $z^{1/2} = \exp[(\frac{1}{2})\text{Log } z]\exp(k\pi i)$, $k = 0, 1$. That is, for $z \neq 0$, $f(z)$ has the two values

$$z^{1/2} = \begin{cases} \exp[(\frac{1}{2})\text{Log } z] = |z|^{1/2} \exp\left[\left(\frac{i}{2}\right)\text{Arg } z\right] \\ \exp[(\frac{1}{2})\text{Log } z]\exp(\pi i) = -|z|^{1/2} \exp\left[\left(\frac{i}{2}\right)\text{Arg } z\right]. \end{cases}$$

Case III   a is an irrational number

Again we write

$$z^a = \exp(a \log z) = \exp(a \text{ Log } z)\exp(2ak\pi i),$$

for $k = 0, \pm 1, \pm 2, \ldots$, and if a is irrational the second factor, $\exp(2ak\pi i)$, has distinct values for distinct values of k. (See problem 2.21.) Then z^a has infinitely many distinct values (one for each value of log z).

If a is now a complex number, we can examine the values of z^a in the following way. For any complex number $a = \alpha + i\beta$, α and β real, we write $z^a = z^\alpha z^{i\beta}$, where z^α is accounted for in one of the cases above. The term $z^{i\beta} = \exp(i\beta \log z) = \exp(-\beta \arg z)\exp(i\beta \log |z|)$ always has infinitely many values corresponding to the values of arg z.

Inverse Trigonometric and Hyperbolic Functions

To return to the paragraph at the beginning of this section, we seek to define a function $w = f(z)$ with the property that for each z, sin w = $\sin[f(z)] = z$. It is not difficult to use the definition of the sine function to solve formally the equation sin $w = z$, as we see below.
    We write

$$(1/2i)(e^{iw} - e^{-iw}) = z, \quad e^{2iw} - 2ize^{iw} - 1 = 0,$$

so that

$$e^{iw} = iz + (1 - z^2)^{1/2},$$

or

$$w = -i \log[iz + (1 - z^2)^{1/2}].$$

For each z such that $iz + (1 - z^2)^{1/2} \neq 0$, we can choose a value for $(1 - z^2)^{1/2}$, and, corresponding to the value of $iz + (1 - z^2)^{1/2}$ so determined, we can select a value for $-i \log[iz + (1 - z^2)^{1/2}]$. For each value of w we obtain, the equation sin $w = z$ will be satisfied. Furthermore, wherever dw/dz exists we have $d(\sin w)/dz = 1$, or $(\cos w)(dw/dz) = 1$, and $dw/dz = \sec w = 1/(1 - z^2)^{1/2}$ — a familiar formula from calculus.
    Without going into detail about why a particular choice is actually made, let us say we can choose a branch of $(1 - z^2)^{1/2}$ which is positive for $-1 < z < 1$ and is analytic for all z except those real z with $|z| \geq 1$. Corresponding to this choice of a branch of $(1 - z^2)^{1/2}$, let us take the principal branch of the logarithm function — that is, the branch for which log $1 = 0$. This determines a principal branch for the inverse sine function which is analytic for all z, except for $z \geq 1$ or $z \leq -1$. It also agrees with the ordinary principal inverse sine function from elementary calculus for $-1 < z < 1$. Denote any branch of the inverse sine function by $w = \sin^{-1}z$.

In order to solve the equation $\cos w = z$ for w, we use the fact that $z = \cos w = \sin(\pi/2 - w)$ to obtain $w = \pi/2 - \sin^{-1} z$. Thus we define the inverse cosine function by $w = \cos^{-1} z = \pi/2 - \sin^{-1} z$, with principal branch determined by the principal branch of $\sin^{-1} z$. Also $\cos^{-1} z$ is differentiable where $\sin^{-1} z$ is differentiable, and $d(\cos^{-1} z)/dz = -1/(1 - z^2)^{1/2}$.

Similarly, if we define the inverse tangent function through solutions for w to the equation $z = \tan w$, we obtain

$$w = \tan^{-1} z = \left(\frac{i}{2}\right) \log\left(\frac{1 - iz}{1 + iz}\right)$$

where a principal branch is specified by taking the principal branch of the logarithm function. Moreover, each branch of $\tan^{-1}z$ is analytic for all z except $z = iy$, where y is real and $|y| \geq 1$, and $d(\tan^{-1} z)/dz = 1/(1 + z^2)$.

Let us summarize our definitions of the inverse trigonometric functions in the following way:

1. Each branch of $\sin^{-1} z = -i \log[iz + (1 - z^2)^{1/2}]$ is analytic for all z, except for $z \geq 1$ or $z \leq -1$, and for these values of z,

    $$\frac{d(\sin^{-1} z)}{dz} = \frac{1}{(1 - z^2)^{1/2}}$$

    We determine a principal branch of $\sin^{-1} z$ by choosing: the branch of $(1 - z^2)^{1/2}$ which is positive for $-1 < z < 1$; and the principal branch of the logarithm function.

2. Each branch of $\cos^{-1} z = \pi/2 - \sin^{-1} z$ is analytic for all z, except for $z \geq 1$ or $z \leq -1$, and for these values of z

    $$\frac{d(\cos^{-1} z)}{dz} = \frac{-1}{(1 - z^2)^{1/2}}$$

    A principal branch of $\cos^{-1} z$ is the branch corresponding to a principal branch of $\sin^{-1} z$.

3. Each branch of $\tan^{-1} z = (i/2)\log\left(\dfrac{1 - iz}{1 + iz}\right)$ is analytic for all z, except for $z = iy$, y real and $|y| \geq 1$, and for all such z,

    $$\frac{d(\tan^{-1} z)}{dz} = \frac{1}{(1 + z^2)}$$

    A principal branch for $\tan^{-1} z$ is determined by the principal branch of the logarithm function.

The remaining three inverse functions can be defined in terms of the three we already have.

The details of arriving at definitions for inverse hyperbolic functions are very similar to those we have discussed for the inverse trigonometric functions. For this reason it is enough to summarize the resulting definitions. The reader might verify for himself some of the formulas involved.

4. Each branch of $\sinh^{-1} z = \log[z + (z^2 + 1)^{1/2}]$ is analytic for all z, except for $z = iy$, where y is real and $|y| \geq 1$, and for all such z,

$$\frac{d(\sinh^{-1} z)}{dz} = \frac{1}{(1 + z^2)^{1/2}}$$

A principal branch of $\sinh^{-1} z$ is determined by choosing: the branch of $(z^2 + 1)^{1/2}$ with value 1 at $z = 0$; and the principal branch of the logarithm function.

5. Each branch of $\cosh^{-1} z = \pi i/2 - \sinh^{-1}(iz)$ is analytic for all z, except for $z \leq -1, z \geq 1$, and for all such z,

$$\frac{d(\cosh^{-1} z)}{dz} = \frac{-1}{(z^2 - 1)^{1/2}}$$

A principal branch of $\cosh^{-1} z$ is the branch corresponding to a principal branch of $\sinh^{-1}(iz)$.

6. Each branch of $\tanh^{-1} z = (\tfrac{1}{2}) \log\left(\dfrac{1 + z}{1 - z}\right)$ is analytic for all z, except for $z \leq -1, z \geq 1$, and for all such z,

$$\frac{d(\tanh^{-1} z)}{dz} = \frac{1}{(1 - z^2)}$$

A principal branch for $\tanh^{-1} z$ is the branch corresponding to the principal branch of the logarithm function.

PROBLEMS

2.18 Find all solutions to the equations:
   (a) $e^z = 17$
   (b) $e^z = 3 \sinh z$

2.19 What is the value of $\mathrm{Log}(1 - i)$?

2.20 Find all the values of
   (a) $i^{1/2}$
   (b) $(1 + i)^{1/4}$
   (c) $8^{1/3}$
   (d) i^i

2.21 If a is an irrational number and k_1, k_2 are distinct integers, show that $\exp(2\pi a k_1 i) \neq \exp(2\pi a k_2 i)$.

2.22 If a is any nonzero constant, let $a^z = \exp(z \log a)$. Where is a^z an analytic function of z, and what is its derivative?

2.23 Verify that whenever the branches of $\cosh^{-1} z$ are analytic, $d(\cosh^{-1} z)/dz = -1/(z^2 - 1)^{1/2}$.

2.24 Verify that whenever the branches of $\tan^{-1} z$ are analytic, $d(\tan^{-1} z)/dz = 1/(1 + z^2)$.

2.25 Verify that

$$\tanh^{-1} z = (\tfrac{1}{2}) \log\left(\frac{1 + z}{1 - z}\right)$$

and that wherever each branch of $\tanh^{-1} z$ is analytic,

$$d(\tanh^{-1} z)/dz = 1/(1 - z^2).$$

2.26 (a) Use problem 1.33 to show that $u(r,\theta) = r^{1/2}\cos(\theta/2)$ is harmonic in the domain $D = \{(r,\theta):1 < r < 2\}$.
   (b) Show there is no function $f(z)$ analytic in D such that $\operatorname{Re} f(z) = u$ everywhere in D.

Section 2.4   Additional Comments on Multiple-Valued Functions

Our treatment of multiple-valued functions in Section 2.3 has been deliberately limited in scope, and in closing this chapter we should like to comment further on this subject.

   In the development of the logarithm function and its branches we quite arbitrarily chose and used the principal value of arg z so that real numbers greater than zero would have principal argument zero and their principal logarithms would coincide with the familiar natural logarithms. While this is frequently most convenient, sometimes it is not. Suppose $f(z)$ is an analytic function in a region D of the complex plane.

If the range of $w = f(z)$ over D contains no points $w \leq 0$, then each branch of $\log f(z)$ is also analytic in D. (Why?) But what if the range of $w = f(z)$ over D does contain points $w \leq 0$? Then we cannot claim that any of the branches we defined for $\log f(z)$ is analytic in all of D.

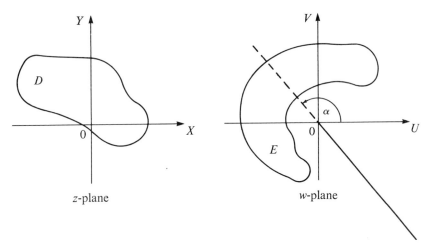

Figure 2.1

Let us suppose that E is the range of $f(z)$ over D and that E contains no points $\{re^{i\alpha} : r \leq 0, \alpha \text{ fixed}\}$ of some ray at $w = 0$ in the w-plane. For any nonzero w we can express the arguments of w, relative to α, by

$$\alpha + (2k - 1)\pi < \arg w \leq \alpha + (2k + 1)\pi, \qquad k \text{ any integer.}$$

And with these values of $\arg w$ we can define branches of the logarithm function for each integer k:

$$\log w = \log |w| + i \arg w, \qquad w \neq 0$$
$$\alpha + (2k - 1)\pi < \arg w \leq \alpha + (2k + 1)\pi$$

which are analytic everywhere except on the ray $\{re^{i\alpha} : r \leq 0\}$. Now the advantage is obvious: each branch of $\log f(z)$ (relative to α) is now analytic in D. Depending on the actual nature of $f(z)$, some branch of $\log f(z)$, and some corresponding value of k, may appear most convenient and so be designated as a "principal" branch of $\log f(z)$.

Since all of Section 2.3 derives from the treatment of the logarithm function, all the functions presented there can be similarly redefined so that we can consider whether branches of $[f(z)]^{1/2}$ or $\sin^{-1}[f(z)]$, for example, are analytic in D.

Whether such a ray $\{re^{i\alpha}:r \leq 0\}$ exists depends on the function $f(z)$ and its range E. Should E contain a region $\{w:0 < r < |w| < s\}$, there would be no way of choosing branches of $\log f(z)$ so as to be analytic in all of D.

However, for any real number α, say $0 \leq \alpha < 2\pi$, we can define branches of $\log z$ relative to α, for $z \neq 0$, by:

$$\log z = \log |z| + i \arg z, \qquad z \neq 0$$

$$\alpha + (2k - 1)\pi < \arg z \leq \alpha + (2k + 1)\pi, \qquad k = 0, \pm 1, \pm 2, \ldots .$$

PROBLEM

2.27 Suppose α is some real number and P is the ray through $z = 0$ described by $\{re^{i\alpha}:r \leq 0\}$. If z_0 is any point not on P, not on the negative real axis, and not zero, are the values of $\log z_0$ relative to α the same as the values of $\log z_0$ as defined in Section 2.3 (which correspond to the value $\alpha = 0$)?

3

Complex Integration

Integration of complex-valued functions has to play a strong and central role in any brief introductory course in functions of a complex variable, and it might be helpful to indicate why this is so before we actually begin our discussion of integration. In the first place, since a function of a complex variable is easily regarded as a vector-valued function of two real variables, the natural form of integration for such functions is line integration. This in turn implies that the elementary calculus emphasis on "antidifferentiation" as a form of integration is less important. And when we try to use information from Green's Theorem in the plane to determine which functions have independent-of-path line integrals, we are immediately led to consider analytic functions again. At this point the power of the classical integration theory, starting with the Cauchy Integral Theorem, begins to tell us a great deal more about analytic functions than we could realize from the techniques of Chapter 1. Finally, in Chapter 5, this information will help us to evaluate improper integrals of real-valued functions.

Section 3.1   Integration of Complex-Valued Functions Along Curves

In order to integrate complex-valued functions along curves in the complex plane, we need to agree on a reasonably precise definition of the word "curve" which fits our intuitive ideas of what the word should mean. One way to define the word is to say that a curve is a continuous complex-valued function defined on a closed bounded interval of the real axis. This rather permissive definition allows us to have many curves representing the same set of points in the complex plane. For our purposes we will find it more convenient to have a definition which associates a curve more closely with a set of points in the plane, so that we can define the integral of a continuous function along the curve independently of the manner in which we describe the curve by a function. Beyond this, we should like our curves to be "smooth" in a geometric sense.

Suppose $x(t)$ and $y(t)$ are continuous real-valued functions of a real variable t for $a \leq t \leq b$, and $z(t) = x(t) + i\, y(t)$. If $x(t)$ and $y(t)$ are differentiable at $t = t_0$, $a < t_0 < b$, define $z'(t_0) = x'(t_0) + i\, y'(t_0)$. We know that $z(t)$ describes a curve in the plane as t goes from a to b, and $z'(t_0)$ is a tangent vector to this curve at the point corresponding to $t = t_0$ as long as $z'(t_0) \neq 0$. We shall call the continuous function $z(t)$ *piecewise smooth* for $a \leq t \leq b$ if: $z'(t)$ is defined and nonzero for all but a finite number of values of t, $a < t < b$; and wherever $z'(t)$ is defined it is continuous. Curves described by piecewise smooth functions will have continuously turning tangent vectors at all but a finite number of points.

Definition   Let $z(t)$ be a complex-valued function, defined and continuous for $0 \leq t \leq 1$, such that
   (1) $z(t)$ is piecewise smooth on $0 \leq t \leq 1$
   (2) $z(t_1) = z(t_2)$ for $0 \leq t_1 \leq t_2 < 1$ if and only if $t_1 = t_2$. The range C of $z(t)$ is called the *path described by* $z(t)$. We say that $z_0 = z(0)$ is the *initial point* of C, $z_1 = z(1)$ is the *terminal point* of C, and C is a path from z_0 to z_1. If $z_0 = z_1$, we call C a *closed path*.

The phrase *curve described by* $z(t)$ is defined by deleting condition (2) above and replacing the word "path" with "curve."

Condition (1) restricts us to paths and curves of finite length, which are smooth except for a finite number of "corners." Condition (2) tells us that as t goes from 0 to 1, $z(t)$ never passes through the same value twice, unless it happens that $z(0) = z(1)$. Actually, conditions (1) and (2) do somewhat more; they allow us to concentrate more on the path described by $z(t)$ than on $z(t)$ itself. Of course, we must know the "tracing order" determined by $z(t)$, but this is so often a tacitly understood part of any context in which paths are useful that we can afford to be somewhat less than explicit about it. In any case, if we have a path C described by $z(t)$, the integrals we define on C in the next several paragraphs are independent of $z(t)$ in the following sense: If $w(t)$ is any other complex-valued function, defined for $0 \leq t \leq 1$, satisfying conditions (1) and (2) and having C as its range, and which traces C in the same direction as $z(t)$, then $z(t)$ can be replaced by $w(t)$ in these integrals without changing their values. (We use these results to partially justify our frequent use of the term "path," which leaves to the context the defining function $z(t)$. Those who wish to read a detailed discussion of these matters are referred to [Apostol].)

If C is a curve from z_0 to z_1, described by a function $z(t)$, and $f(z) = u(x,y) + iv(x,y)$ is defined at every point of C, on a formal basis we may write

$$
\begin{aligned}
f(z)\,dz &= f[z(t)]\,dz(t) \\
&= \{u[x(t),y(t)] + iv[x(t),y(t)]\}[dx(t) + i\,dy(t)] \\
&= \{u[x(t),y(t)] + iv[x(t),y(t)]\}[x'(t) + i\,y'(t)]\,dt \\
&= \{u[x(t),y(t)]x'(t) - v[x(t),y(t)]\,y'(t)\}\,dt \\
&\quad + i\{v[x(t),y(t)]x'(t) + u[x(t),y(t)]\,y'(t)\}\,dt
\end{aligned}
$$

wherever $x'(t)$ and $y'(t)$ are defined. As t goes from 0 to 1, $z(t)$ goes from z_0 to z_1 along C, and it is reasonable to take the integral of $f(z)$ from z_0 to z_1 along C to be

$$
\int_0^1 \{u[x(t),y(t)]x'(t) - v[x(t),y(t)]y'(t)\}\,dt
$$

$$
+ i\int_0^1 \{v[x(t),y(t)]x'(t) + u[x(t),y(t)y'(t)]\}\,dt
$$

To take account of the piecewise smooth nature of $z(t)$, each of these integrals should be replaced by a finite sum of integrals. We will so understand but continue to use the brief versions. The existence of these integrals then depends solely on $f(z)$; the assumption that $f(z)$ be continuous on C will surely be enough to guarantee that these integrals do exist.

For our purposes it is enough to consider integrals of continuous functions, so we make the suggestions above a formal definition.

Definition Let z_0 and z_1 be any points in the complex plane and let C be a curve from z_0 to z_1, described by $z(t)$, $0 \leq t \leq 1$. If $f(z) = u(x,y) + iv(x,y)$ is a function continuous at every point of C, then the *integral of* $f(z)$ *from* z_0 *to* z_1 *along* C is defined to be

$$\int_{z_0}^{z_1} f(z)dz = \int_0^1 \{u[x(t),y(t)]x'(t) - v[x(t),y(t)]y'(t)\} dt$$

$$+ i \int_0^1 \{v[x(t),y(t)]x'(t) + u[x(t),y(t)]y'(t)\} dt$$

where $z(t) = x(t) + iy(t)$, $0 \leq t \leq 1$.

The hypotheses on $f(z)$ and C are enough to ensure that $\int_{z_0}^{z_1} f(z)\,dz$ does exist. Furthermore, if C is a path described by $z(t)$, the value of $\int_{z_0}^{z_1} f(z)\,dz$ is the same for all functions $w(t)$ describing C which satisfy conditions (1) and (2) and trace C in the same direction as $z(t)$.

It is also convenient to write the equation above in a familiar shorthand, where $z = x + iy$ lies on C, and $f(z) = u + iv$ gives the values of $f(z)$ on C:

$$\int_{z_0}^{z_1} f(z)\,dz = \int_{(x_0,y_0)}^{(x_1,y_1)} [udx - vdy] + i \int_{(x_0,y_0)}^{(x_1,y_1)} [vdx + udy]$$

Example 3.1 Evaluate $\int_0^{1+i} e^z\,dz$, where C is the straight line path from 0 to $1 + i$ described by the function $z(t) = t + it$, $0 \leq t \leq 1$.

$$\int_0^{1+i} e^z\,dz = \int_{(0,0)}^{(1,1)} [e^x \cos y\,dx - e^x \sin y\,dy]$$

$$+ i \int_{(0,0)}^{(1,1)} [e^x \sin y\,dx + e^x \cos y\,dy]$$

$$= \int_0^1 (e^t \cos t - e^t \sin t) \, dt$$

$$+ i \int_0^1 (e^t \sin t + e^t \cos t) \, dt$$

$$= e(\cos 1 + i \sin 1) - 1 = \exp(1 + i) - 1.$$

Note that e^z is the derivative of a function, e^z, analytic on C, and evaluating this integral as if "antidifferentiation" were legal would have given us $\exp(1 + i) - \exp 0 = \exp(1 + i) - 1$. If this is not coincidence, we should like to know it, for evaluating line integrals directly can be tedious.

Example 3.2   Let C_1 be the path from $-i$ to i along the right semicircle $\{z : |z| = 1, \text{ Re } z \geq 0\}$ and C_2 be the path from $-i$ to i along the left semicircle $\{z : |z| = 1, \text{ Re } z \leq 0\}$. We wish to evaluate the integrals

$$\int_{\substack{-i \\ C_j}}^{i} \frac{dz}{z}, \text{ for } j = 1, 2.$$

On C_1, $z = e^{i\theta}$, $-\pi/2 \leq \theta \leq \pi/2$, and $dz = ie^{i\theta} \, d\theta$, so

$$\int_{\substack{-i \\ C_1}}^{i} \frac{dz}{z} = i \int_{-\pi/2}^{\pi/2} d\theta = i\pi.$$

Similarly

$$\int_{\substack{-i \\ C_2}}^{i} \frac{dz}{z} = -i\pi.$$

Clearly integrals of $1/z$ are not always independent of path. Also $1/z$ is the derivative of any branch of the logarithm function for any nonzero z. If we try to evaluate these integrals by "antidifferentiation," which branches of $\log z$ should we use corresponding to C_1 and C_2 in the example above?

The basic properties for the integral we have defined will follow directly from familiar properties of line integrals of vector functions.

P1 If $f(z)$ and $g(z)$ are continuous at each point of a curve C from z_0 to z_1 described by $z(t)$, $0 \leq t \leq 1$, and if k and m are any fixed complex numbers, then

$$\int_{z_0}^{z_1} [kf(z) + mg(z)] \, dz = k \int_{z_0}^{z_1} f(z) \, dz + m \int_{z_0}^{z_1} g(z) \, dz$$
$$C \qquad\qquad\qquad\qquad\qquad C \qquad\qquad C$$

P2 If $f(z)$ is continuous at each point of a curve C from z_0 to z_1 described by $z(t)$, $0 \leq t \leq 1$, and we let $-C$ denote the curve from z_1 to z_0 described by $z(1 - t)$, $0 \leq t \leq 1$, then

$$\int_{z_1}^{z_0} f(z) \, dz = -\int_{z_0}^{z_1} f(z) \, dz$$
$$-C \qquad\qquad\qquad C$$

P3 Let C_1 be a curve from z_0 to z_1 described by $z_1(t)$ and C_2 be a curve from z_1 to z_2 described by $z_2(t)$. If we define

$$z(t) = \begin{cases} z_1(2t), & 0 \leq t \leq \tfrac{1}{2} \\ z_2(2t - 1), & \tfrac{1}{2} \leq t \leq 1 \end{cases}$$

then $C = \{z(t) : 0 \leq t \leq 1\}$ is a curve from z_0 to z_2 which we call the sum of C_1 and C_2 and write $C = C_1 \oplus C_2$. If $f(z)$ is continuous at each point of C_1 and C_2, then

$$\int_{z_0}^{z_2} f(z) \, dz = \int_{z_0}^{z_1} f(z) \, dz + \int_{z_1}^{z_2} f(z) \, dz$$
$$C \qquad\qquad\qquad C_1 \qquad\qquad\qquad C_2$$

P4 If C is a curve from z_0 to z_1 described by $z(t) = x(t) + i\, y(t)$, then

$$\int_{z_0}^{z_1} |dz| = \int_0^1 [x'(t)^2 + y'(t)^2]^{1/2} \, dt$$
$$C$$

$$= \text{length along } C \text{ from } z_0 \text{ to } z_1$$

P5 If C is a curve from z_0 to z_1, with $\int_{z_0}^{z_1} |dz| = L$, and $f(z)$ is continuous at each point of C, then
$$C$$

$$\left| \int_{z_0 \atop C}^{z_1} f(z)\, dz \right| \le \int_{z_0 \atop C}^{z_1} |f(z)||dz| \le ML$$

where M is any constant such that $|f(z)| \le M$ for all z on C.

Example 3.3   Let C be the polygonal path from 0 to $1 + i$ consisting of the line segment C_1 from 0 to 1 and the line segment C_2 from 1 to $1 + i$. We can write

$$C_1 = \{z_1(t) = t : 0 \le t \le 1\}$$

and

$$C_2 = \{z_2(t) = 1 + it : 0 \le t \le 1\},$$

so that $C = C_1 \oplus C_2 = \{z(t) : 0 \le t \le 1\}$, where

$$z(t) = \begin{cases} z_1(2t) = 2t & 0 \le t \le \tfrac{1}{2} \\ z_2(2t - 1) = 1 + i(2t - 1) & \tfrac{1}{2} \le t \le 1 \end{cases}$$

From P3 we have

$$\int_{0 \atop C}^{1+i} \sin z\, dz = \int_{0 \atop C_1}^{1} \sin z\, dz + \int_{1 \atop C_2}^{1+i} \sin z\, dz$$

$$= \int_0^1 \sin t\, dt + i \int_0^1 \sin(1 + it)\, dt$$

$$= 1 - \cos 1$$

$$+ i \int_0^1 (\sin 1 \cosh t + i \cos 1 \sinh t\, dt)$$

$$= 1 - \cos 1 \cosh 1 + i \sin 1 \sinh 1$$

$$= 1 - \cos(1 + i).$$

As we progress through this chapter we shall come to work primarily with integrals about closed paths. We adopt a familiar convention for distinguishing between positively and negatively oriented closed paths.

Definition   Let C be a closed path described by $z(t)$. If, as t goes from 0 to 1, C is traced out in such a way that points enclosed by C are always on the left (respectively, right), C is said to be *positively oriented* (respectively, *negatively oriented.*)

Unless otherwise indicated, when we refer to a closed path C, it will be understood that C is positively oriented. To denote the same path with a negative orientation, we shall write $-C$.

PROBLEMS

3.1   If C is the boundary of a rectangle with vertices $1 + i$, $-1 + i$, $-1 - i$, and $1 - i$, and if $f(z) = z^2$, evaluate

$$\int_C f(z)\, dz.$$

3.2   Let $z = a$ and $z = b$ be two fixed points and C be any curve from a to b. Show that

$$\int_{C\,a}^{\;\;b} z\, dz = \frac{(b^2 - a^2)}{2}.$$

3.3   If C is the straight line path from 0 to $1 + i$, is it true that

$$\int_{C\,0}^{\;\;1+i} \cos z\, dz = \sin(1 + i)?$$

3.4   If a is any fixed point and C is the circle described by $|z - a| = 2$, and if n is any integer, show that

$$\int_C \frac{dz}{(z - a)^n} = \begin{cases} 0 \text{ for } n \neq 1 \\ 2\pi i \text{ for } n = 1 \end{cases}$$

3.5   Let C be a circle, $|z| = R > 0$, and $P(z) = a_0 + a_1 z + a_2 z^2 + \cdots + a_n z^n$ be any polynomial of degree n, where n is a positive integer. Use the previous problem to show that $\int_C P(z)\, dz = 0$.

3.6   If C is the circle $|z| = 2$, evaluate $\int_C |z - 2||dz|$.

3.7   Let C be the path $\{z : |z| = 1, \text{Re } z \geq 0\}$ from $-i$ to i. Show that $\left| \int_C (x^2 + iy^2) \, dz \right| \leq \pi$.

Section 3.2   Functions Defined by Indefinite Integrals

Suppose D is a simply connected domain in the complex plane and $f(z)$ is continuous in D. Let a and b be points of D and C be a path in D from a to b. From Example 3.1 we can make a cautious guess that if $F(z)$ is analytic in D and $F'(z) = f(z)$ at every point of D, then

$$\int_{\substack{a \\ C}}^{b} f(z) \, dz = F(b) - F(a).$$

The value that this guess assigns to $\int_{\substack{a \\ C}}^{b} f(z) \, dz$ does not depend on C.

If for every pair of points, a and b, in D and any path C in D from a to b, the integral $\int_{\substack{a \\ C}}^{b} f(z) dz$ is a number depending only on a and b and not on the path C, then we say the integral $\int_{\substack{a \\ C}}^{b} f(z) \, dz$ is *independent of path* from a to b in D, and we write its value as $\int_a^b f(z) \, dz$.

If we keep a fixed but let $b = w$ be any point of D and suppose that $\int_a^w f(z) dz$ has meaning without reference to a path in D for each w in D, then the integral $\int_a^w f(z) \, dz$ defines a function of w in D. If we write

$$F(w) = \int_a^w f(z) \, dz,$$

elementary calculus would suggest that $F'(w) = f(w)$.

To collect these hunches together and assert their accuracy, we state a theorem and give an outline of its proof.

THEOREM 3.1   Hypotheses:
H1 D is a simply connected domain.
H2 $f(z)$ is continuous in D.

Conclusions:
C1   For every pair of points, a and b, in D, and any C in D from a to b,
$\int_a^b f(z)\,dz$ is independent of path in D if and only if there exists a
function $F(z)$, analytic in D, such that $f(z) = F'(z)$ at every point
of D.
C2   If such a function $F(z)$ exists,

$$\int_a^b f(z)\,dz = F(b) - F(a).$$

C3   If, for a fixed point a of D, $\int_a^w f(z)\,dz = F(w)$ is defined and independent of path in D for any w in D, then $F(w)$ is analytic in D,
and $F'(w) = f(w)$ for each w in D.

Outline of proof: If $\int_a^b f(z)\,dz$ is independent of path in D and $f = u + iv$, then

$$\int_a^b f(z)\,dz = \int_a^b [u\,dx - v\,dy] + i\int_a^b [v\,dx + u\,dy],$$

where both these last integrals are independent of path in D. Then there
exist real-valued functions U and V in D such that $U_x = u$, $U_y = -v$,
$V_x = v$, and $V_y = u$ everywhere in D, and

$$\int_a^b [u\,dx - v\,dy] = U(b) - U(a),\ \int_a^b [v\,dx + u\,dy] = V(b) - V(a).$$

The function $F(z) = U(z) + i\,V(z)$ satisfies all the hypotheses of Theorem
1.5, so $F(z)$ is analytic in D, and $F'(z) = f(z)$. Also

$$\int_a^b f(z)\,dz = F(b) - F(a).$$

If there is a function $F(z)$, analytic in D with $F'(z) = f(z)$, let $F(z) = U(z) + i\,V(z)$. Then, Theorem 1.4, $U_x = u = V_y$, and $V_x = v = -U_y$. For any path C in D from a to b,

$$\int_{\substack{a\\C}}^{b} f(z)\,dz = \int_{\substack{a\\C}}^{b} [u\,dx - v\,dy] + i\int_{\substack{a\\C}}^{b} [v\,dx + u\,dy]$$

$$= \int_{\substack{a\\C}}^{b} dU + i\int_{\substack{a\\C}}^{b} dV$$

$$= U(b) - U(a) + i\,[V(b) - V(a)]$$

$$= F(b) - F(a)$$

which is independent of the path C.

If $f(z)$ is continuous in D and is the derivative of some function $F(z)$ analytic in D, then for any fixed point z_0 in D we call $F(z;z_0) = \int_{z_0}^{z} f(w)\,dw$ an *indefinite integral* of $f(z)$ in D. We can see that for any a and b of D,

$$\int_{a}^{b} f(w)\,dw = \int_{z_0}^{b} f(w)\,dw - \int_{z_0}^{a} f(w)\,dw$$

$$= F(b;z_0) - F(a;z_0).$$

Since for each z_0 in D the function $F(z;z_0)$ must have $f(z)$ as its derivative at each point of D, the only effect on $F(z;z_0)$ of changing z_0 is the adding of some constant. (Proof?)

Example 3.4 If a and b are any two complex numbers and n is a positive integer, let $f(z) = z^n$. Since $f(z)$ is continuous for all z and is the derivative of $F(z) = z^{n+1}/(n+1)$, a function analytic for all z, we know

$$\int_{a}^{b} z^n\,dz = \frac{(b^{n+1} - a^{n+1})}{(n+1)}.$$

Example 3.5 Again let a and b be any two distinct, nonzero complex numbers, and let $f(z) = 1/z$. Now $f(z)$ is continuous for all nonzero z, and each branch of the logarithm function has derivative $1/z$ for all those values of z where it is analytic.

69438

Let C be any path from a to b not passing through $z = 0$ and for which we can choose a ray $A = \{z : |z| \geq 0, \arg z = \alpha\}$, for some real number α, which does not intersect C. We can determine the branches of the logarithm function so as to be analytic everywhere except for points on A. (See Fig. 3.1.)

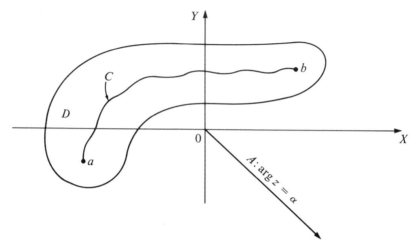

Figure 3.1

If $F(z) = \log z$ represents any such branch, then $F'(z) = 1/z$, and all the hypotheses of Theorem 3.1 are satisfied for any simply connected domain D containing C but excluding A. Thus

$$\int_{a \atop C}^{b} \frac{dz}{z} = \int_a^b \frac{dz}{z} = \log b - \log a.$$

Since any two branches of $\log z$ which are analytic except for the points on A differ by only a constant, if we let Log z represent a principal branch of the logarithm function, analytic except for points on A, then

$$\int_a^b \frac{dz}{z} = \text{Log } b - \text{Log } a$$

as well. There may be many possible choices for a ray A not intersecting C; the import of problem 2.27 is that the right side of the preceding equation will remain the same for all such choices of A. Finally we can repeat this process for any path C from a to b not passing through $z = 0$ for which at least one choice for A exists. (That is, there is at least one direction we can travel from $z = 0$ in a straight line path without crossing C.) Thus we may conclude that for any distinct nonzero numbers a and b,

$$\int_a^b \frac{dz}{z} = \text{Log } b - \text{Log } a,$$

the result being independent of path for any simply connected domain in the plane containing a and b but excluding some ray A.

PROBLEMS

3.8 Evaluate $\int_0^i (z^2 - 2z + 1)\, dz$.

3.9 Evaluate $\int_0^{k\pi i} e^z\, dz$ for any integer k.

3.10 Evaluate $\int_{C}^{a} \text{Log } z\ dz$, where a is any number with Im $a \neq 0$.

Here $\text{Log } z = \text{Log } |z| + i \text{ Arg } z$, and C is any path from 1 to a not intersecting the ray $\{z : |z| \geq 0, \arg z = -\pi\}$.

3.11 If a is any positive real number, determine where $\int_1^a z^{1/2}\, dz$ is independent of path and give its value.

Section 3.3 The Cauchy Integral Theorem

The Cauchy Integral Theorem seems rather modest when we actually state it. In fact, with the hypotheses we use and the simplest form of the theorem which we state, the theorem and its proof do not appear to be very deep. This apparent simplicity is quite misleading, for the theorem is a most powerful result and provides the basis for almost all the discussions which follow. The relatively narrow statement and proof we present are tailored to our needs. Instead of trying to describe in advance the important role this theorem plays, we let the theorem speak for itself through the uses we make of it. We give only a partial proof of the theorem we state, for we shall add a hypothesis which is actually re-dundant — but not obviously so — and assume that the closed curve in H3 is a path. We refer the reader to [Pennisi] for a more complete discussion.

THEOREM 3.2 The Cauchy Integral Theorem
Hypotheses:
H1 D is a simply-connected domain.
H2 $f(z)$ is analytic in D.
H3 C is any closed curve in D.

Conclusion: $\int_C f(z)\, dz = 0.$

Partial Proof: We prove the theorem under the additional hypotheses that $f'(z)$ is continuous in D, and C is a closed path. (The first of these additional hypotheses makes the proof very simple, but it is redundant; it can be proved that if $f(z)$ is analytic in D, so is $f'(z)$, and $f'(z)$ is certainly continuous in D.)
    If

$$f(z) = u(x,y) + iv(x,y),$$

then

$$f'(z) = u_x(x,y) + iv_x(x,y) = v_y(x,y) - iu_y(x,y)$$

by the Cauchy-Riemann equations. The hypothesis we have placed on $f'(z)$ implies that u_x, v_x, u_y, v_y are all continuous in D.
    If T denotes the domain interior to C, we can use Green's Theorem in the plane to write

$$\int_C f(z)\, dz = \int_C [u\,dx - v\,dy] + i \int_C [v\,dx + u\,dy]$$

$$= \iint_T (-v_x - u_y)\, dx\, dy$$

$$+ i \iint_T (u_x - v_y)\, dx\, dy$$

since all the hypotheses of Green's Theorem for $u(x,y)$ and $v(x,y)$ on C and in T are present here. The Cauchy-Riemann equations allow us to conclude that $\int_C f(z)\,dz = 0$.

Of course, this theorem is related to our discussion of independence of path in the previous section. The second conclusion in the following corollary to the Cauchy Integral Theorem indicates this relation.

Corollary 3.1   Hypotheses:
H1 D is a simply connected domain.
H2 $f(z)$ is analytic in D.

Conclusions:
C1   For any points a and b of D and any paths C_1 and C_2 in D from a to b,

$$\int_{C_1}^{b} f(z)\,dz = \int_{C_2}^{b} f(z)\,dz.$$

C2   There exists a function $F(z)$, analytic in D, such that $F'(z) = f(z)$ at every point of D.

Proof: Let $C = C_1 \oplus (-C_2)$. C is a closed path in D and, using Theorem 3.2 and properties P2 and P3 of Section 3.1,

$$0 = \int_C f(z)\,dz = \int_{C_1}^{b} f(z)\,dz + \int_{-C_2}^{a} f(z)\,dz$$

$$= \int_{C_1}^{b} f(z)\,dz - \int_{C_2}^{b} f(z)\,dz$$

C2 now follows from C1 by Theorem 3.1.

The next theorem we state helps us in both Section 3.4 and Chapter 4. It, too, deals with independence of path for integrals about closed paths and contains Theorem 3.2 as a special case. As do most results on independence of path, this theorem gives conditions under which the integral of a function along a closed path will be independent of the "shape" of the closed path.

THEOREM 3.3   Hypotheses:

H1 a is a fixed point and $0 \leq R < S < T$.

H2 D is the domain consisting of all points z such that $R < |z - a| < T$.

H3 C_1 is the circle $|z - a| = S$.

H4 $f(z)$ is analytic in D.

H5 C is any closed path lying in D whose interior contains C_1.

Conclusion: $\displaystyle\int_C f(z)\,dz = \int_{C_1} f(z)\,dz.$

Proof: If in fact $f(z)$ is analytic in the domain $\{z : |z| < T\}$, then both integrals above are zero by the Cauchy Integral Theorem.

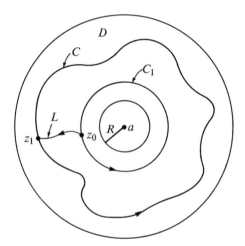

Figure 3.2

For any point z_0 on C_1 construct a path L from z_0 to some point z_1 on C such that L lies inside C and in D except for the point z_1. We see that $f(z)$ is analytic on and inside the closed curve $K = C \oplus (-L) \oplus (-C_1) \oplus L$, so that

$$0 = \int_K f(z)\,dz = \int_C f(z)\,dz + \int_{-L} f(z)\,dz + \int_{-C_1} f(z)\,dz + \int_L f(z)\,dz$$

$$= \int_C f(z)\,dz - \int_L f(z)\,dz - \int_{C_1} f(z)\,dz + \int_L f(z)\,dz$$

and

$$\int_C f(z)\, dz = \int_{C_1} f(z)\, dz.$$

This theorem has an immediate and useful corollary.

Corollary 3.2   Hypotheses:
H1–H4   as in Theorem 3.3
H5 C and K are any closed paths in D whose interiors contain C_1.

Conclusion: $\displaystyle\int_C f(z)\, dz = \int_K f(z)\, dz.$

Example 3.6   For a positive number r, let $C(r)$ be the circle $\{z : |z| = r\}$.
Evaluate $\displaystyle\int_{C(r)} dz/z^n$, where n is any integer.

Any point on $C(r)$ can be written in the form $z = r \exp(i\theta)$, $-\pi \leq \theta \leq \pi$, and

$$\int_{C(r)} \frac{dz}{z^n} = \int_{-\pi}^{\pi} \frac{ir \exp(i\theta)d\theta}{r^n \exp(in\theta)}$$

$$= \left(\frac{i}{r^{n-1}}\right) \int_{-\pi}^{\pi} \exp[-i(n-1)\theta]d\theta.$$

If $n = 1$, obviously

$$\int_{C(r)} \frac{dz}{z} = 2\pi i.$$

For $n \neq 1$ the reader can verify that

$$\int_{-\pi}^{\pi} \exp[-i(n-1)\theta]\, d\theta = 0,$$

so that

$$\int_{C(r)} \frac{dz}{z^n} = \begin{cases} 0 \text{ for } n \neq 1 \\ 2\pi i \text{ for } n = 1 \end{cases}$$

Example 3.7   If C is any closed path not passing through $z = 0$, and n is any positive integer, find all possible values for $\int_C dz/z^n$.

Case I $z = 0$ is outside C.
    In this case $1/z^n$ is analytic in a simply-connected domain containing C, and by Theorem 3.2 $\int_C dz/z^n = 0$.

Case II $z = 0$ is inside C.
    We can choose a positive number r so small that $C(r) = \{z : |z| = r\}$ is contained inside C and

$$\int_C \frac{dz}{z^n} = \int_{C(r)} \frac{dz}{z^n}.$$

From the previous example we see that $\int_C dz/z^n$ is 0 if $n \neq 1$ and $2\pi i$ if $n = 1$.

PROBLEMS

3.12 If a is any fixed point and for a positive number r, $C(r)$ is the circle $\{z : |z - a| = r\}$, evaluate $\int_{C(r)} dz/(z - a)^n$ for each integer n.

3.13 If C is any closed path not passing through $z = a$ and n is an integer, find all possible values for $\int_C dz/(z - a)^n$.

3.14 Let $f(z) = 1/(z^2 - 4z + 3)$. Evaluate $\int_C f(z)\, dz$, where:

(a) C is the circle $\{z : |z| = 4\}$;
(b) C is the circle $\{z : |z| = 2\}$;
(c) C is the circle $\{z : |z - 4| = 2\}$.
(*Hint:*   Use partial fractions to rewrite $f(z)$ in terms of $1/(z - 3)$ and $1/(z - 1)$.)

3.15 Do problem 3.14 again with

$$f(z) = \frac{(4z + 22)}{[(z^2 + 25)(z^2 - 4z + 3)]}$$

3.16 Suppose $f(z)$ is continuous for $|z| < 1$. For each r, $0 < r < 1$, let $C(r)$ be the circle $\{z:|z| = r\}$. If for every r, $0 < r < 1$, we know that $\int_{C(r)} f(z)\,dz = 0$, must $f(z)$ be analytic for $|z| < 1$?

Section 3.4   Consequences of the Cauchy Integral Theorem

The power of the Cauchy Integral Theorem lies in what it can tell us about analytic functions. If the reader were to take stock now of what he knows about analytic functions, his information would be limited to the Cauchy-Riemann equations and the statement of the Cauchy Integral Theorem. A similar stock-taking at the time he finishes the problems of this section should provide a useful measurement for the importance of the Cauchy Integral Theorem.

The first theorem we state gives us a collection of results known as the *Cauchy integral formulas* and at the same time shows us that an analytic function has derivatives of all orders at each point where it is analytic — that is, that the derivative of an analytic function is itself an analytic function.

THEOREM 3.4   Hypotheses:
H1 $f(z)$ is analytic in a simply connected domain D.
H2 C is any closed path in D.
H3 z_0 is any point inside C.

Conclusions:

C1 $f(z_0) = (1/2\pi i)\int_C f(z)/(z - z_0)\,dz$.

C2   For each positive integer n, $f(z)$ has an nth derivative at z_0 given by

$$f^{(n)}(z_0) = \frac{n!}{2\pi i}\int_C \frac{f(z)}{(z - z_0)^{n+1}}dz, \quad n = 1, 2, 3, \ldots.$$

C3   For each positive integer n, $f^{(n)}(z)$ is analytic in D.

Partial Proof: The equations of C1 and C2 together are known as the *Cauchy integral formulas.* Write

$$(1/2\pi i) \int_C \left[\frac{f(z)}{(z-z_0)}\right] dz = (1/2\pi i) \int_C \left\{\frac{[f(z)-f(z_0)]}{(z-z_0)}\right\} dz$$

$$+ (1/2\pi i) \int_C \left[\frac{f(z_0)}{(z-z_0)}\right] dz.$$

This last integral equals $[f(z_0)/2\pi i] \int_C dz/(z-z_0)$, where C is a closed path about $z = z_0$. Using problem 3.13, we have

$$(1/2\pi i) \int_C \left[\frac{f(z)}{(z-z_0)}\right] dz = (1/2\pi i) \int_C \left\{\frac{[f(z)-f(z_0)]}{(z-z_0)}\right\} dz + f(z_0).$$

Now $[f(z) - f(z_0)]/(z - z_0)$ is analytic throughout D except at the point $z = z_0$. If r is any positive number so small that $C(r) = \{z:|z - z_0| = r\}$ is contained inside C, by Theorem 3.3

$$(1/2\pi i) \int_C \left\{\frac{[f(z)-f(z_0)]}{(z-z_0)}\right\} dz = (1/2\pi i) \int_{C(r)} \left\{\frac{[f(z)-f(z_0)]}{(z-z_0)}\right\} dz.$$

Since $f(z)$ is analytic at z_0, it is continuous at z_0; given any number $\epsilon > 0$ we can find a number $\delta(\epsilon,z_0) > 0$ such that $|f(z) - f(z_0)| < \epsilon$ whenever $|z - z_0| < \delta(\epsilon,z_0)$. Now choose any r so that $0 < r < \delta(\epsilon,z_0)$. For this choice of r, $|f(z) - f(z_0)| < \epsilon$ for all z on $C(r)$, and

$$\left|(1/2\pi i) \int_{C(r)} \left\{\frac{[f(z)-f(z_0)]}{(z-z_0)}\right\} dz\right|$$

$$\leq (1/2\pi) \int_{C(r)} \left\{\frac{|f(z)-f(z_0)|}{|z-z_0|}\right\} dz \leq \left(\frac{\epsilon}{2\pi}\right) \int_{C(r)} \left(\frac{1}{r}\right) |dz|$$

$$= \left(\frac{\epsilon}{2\pi r}\right) \int_{C(r)} |dz| = \epsilon.$$

Since $\epsilon > 0$ was arbitrarily chosen, it must be that

$$(1/2\pi i) \int_{\check{C}(r)} \left\{ \frac{[f(z) - f(z_0)]}{(z - z_0)} \right\} dz = 0,$$

and

$$f(z_0) = (1/2\pi i) \int_C \left[\frac{f(z)}{(z - z_0)} \right] dz,$$

proving C1.

To verify C2 for the case $n = 1$, write $f'(z_0) = \lim_{h \to 0} \{[f(z_0 + h) - f(z_0)]/h\}$. We show that we can make

$$\left| \left\{ \frac{[f(z_0 + h) - f(z_0)]}{h} \right\} - (1/2\pi i) \int_C \left[\frac{f(z)}{(z - z_0)^2} \right] dz \right|$$

arbitrarily small by making $|h|$ sufficiently close to 0. This will show that

$$f'(z_0) = \lim_{h \to 0} \left\{ \frac{[f(z_0 + h) - f(z_0)]}{h} \right\}$$

$$= (1/2\pi i) \int_C \left[\frac{f(z)}{(z - z_0)^2} \right] dz.$$

From C1,

$$\frac{[f(z_0 + h) - f(z_0)]}{h} = (1/2\pi i h) \int_C f(z) \left\{ \left[\frac{1}{(z - z_0 - h)} \right] - \left[\frac{1}{(z - z_0)} \right] \right\} dz$$

$$= (1/2\pi i) \int_C \left\{ \frac{f(z)}{[(z - z_0 - h)(z - z_0)]} \right\} dz,$$

and $$\left\{ \frac{[f(z_0 + h) - f(z_0)]}{h} \right\} - (1/2\pi i) \int_C \left[\frac{f(z)}{(z - z_0)^2} \right] dz$$

$$= (1/2\pi i) \int_C f(z) \left\{ \left[\frac{1}{(z - z_0 - h)(z - z_0)} \right] - \left[\frac{1}{(z - z_0)^2} \right] \right\} dz$$

If we choose $r > 0$ so small that $C(r) = \{z : |z - z_0| = r\}$ lies inside C and choose h so that $0 < |h| < (r/2)$, then this last integral equals

$$(1/2\pi i) \int_{\check{C}(r)} f(z) \left\{ \left[\frac{1}{(z - z_0 - h)(z - z_0)} \right] - \left[\frac{1}{(z - z_0)^2} \right] \right\} dz.$$

(Why is this so?)

On $C(r)$, $|f(z)|$ is a continuous function and so has a least upper bound M. On $C(r)$, $z - z_0 = r \exp(i\theta)$, and

$$\left| \left[\frac{1}{(z - z_0 - h)(z - z_0)} \right] - \left[\frac{1}{(z - z_0)^2} \right] \right|$$

$$= \left| \left[\frac{1}{(r \exp(i\theta) - h)(r \exp(i\theta))} \right] - \left[\frac{1}{r^2 \exp(2i\theta)} \right] \right|$$

$$= \left| \frac{hr \exp(i\theta)}{\{r^2 \exp(2i\theta)[r^2 \exp(2i\theta) - hr \exp(i\theta)]\}} \right|$$

$$\leq \frac{|h|}{[r(r^2 - |h|r)]} < \frac{2|h|}{r^3},$$

since $|h| < (r/2)$.

Now, given any $\epsilon > 0$, if we further restrict h so that $|h| < \delta(\epsilon, z_0) = r^2\epsilon/2M$, we find that

$$\left| (1/2\pi i) \int_{\check{C}(r)} f(z) \left\{ \left[\frac{1}{(z - z_0 - h)(z - z_0)} \right] - \left[\frac{1}{(z - z_0)^2} \right] \right\} dz \right|$$

$$\leq (1/2\pi)M \left(\frac{2|h|}{r^3} \right) \int_{\check{C}(r)} |dz|$$

$$< (1/2\pi)M \left(\frac{2|h|}{r^3} \right)(2\pi r) < \epsilon.$$

Consequently, we see that when $|h| < \delta(\epsilon, z_0)$,

$$\left| \left\{ \frac{[f(z_0 + h) - f(z_0)]}{h} \right\} - (1/2\pi i) \int_C \left[\frac{f(z)}{(z - z_0)^2} \right] dz \right| < \epsilon$$

The path has been tortuous to justify C2 for the case $n = 1$. A full proof of C2 can be made by mathematical induction on n, which requires considerable manipulation and deft handling of inequalities.

Once C2 is established for all positive integers n, C3 follows by definition of "analytic" and the fact that for any point z_0 in D we can construct a closed path in D about z_0.

One of the implications of Theorem 3.4 can be phrased this way: If $f(z)$ is analytic on and inside a closed path C, and if we know the values of $f(z)$ at all points of C, then the values of $f(z)$ and all its derivatives are completely determined at all points inside C. However, the important use we make of the Cauchy integral formulas is not in determining function values by evaluating integrals; rather it is in using values of $f(z)$ and its derivatives to evaluate a variety of integrals.

Example 3.8   Let C be the circle $\{z : |z| = 1\}$ and let n be any integer. Evaluate $I(n) = \int_C [\cos z / z^{n+1}]\, dz$.

Since $\cos z$ is analytic for all z, for any integer $n \leq -1$, $I(n) = 0$ by the Cauchy Integral Theorem. For $n \geq 0$ we can use the Cauchy integral formulas to conclude that

$$I(n) = \frac{2\pi i}{n!} \left[\frac{d^n(\cos z)}{dz^n} \right]_{z=0}$$

$$= \begin{cases} 0 & \text{if } n \text{ is odd} \\ \dfrac{(-1)^k 2\pi i}{(2k)!} & \text{if } n = 2k,\ k = 0, 1, 2, \ldots \end{cases}$$

Example 3.9   Let C be any closed path enclosing the points $z = i$, $z = -i$. Evaluate the integral $(1/2\pi i) \int_C [e^z/(z^2 + 1)]\, dz$.

Since

$$\frac{1}{(z^2 + 1)} = (1/2i)\left\{ \left[\frac{1}{(z - i)} \right] - \left[\frac{1}{(z + i)} \right] \right\},$$

$$(1/2\pi i) \int_C \left[\frac{e^z}{(z^2 + 1)} \right] dz$$

$$= (1/2\pi i)\left\{ (1/2i) \int_C \left[\frac{e^z}{(z - i)} \right] dz - (1/2i) \int_C \left[\frac{e^z}{(z + i)} \right] dz \right\}$$

$$= (1/2i)(e^i - e^{-i}) = \sin 1.$$

Example 3.10   Suppose $f(z)$ is analytic in a simply connected domain D and z_0 is a point of D. For any closed path C in D not passing through z_0 we can show that

$$\int_C \left[\frac{f'(z)}{(z - z_0)}\right] dz = \int_C \left[\frac{f(z)}{(z - z_0)^2}\right] dz.$$

In the case where z_0 is outside C, both integrals will be equal to zero. (Why?) If z_0 is inside C the first integral equals $2\pi i\, f'(z_0)$ — by virtue of the Cauchy integral formula for $n = 0$ applied to the function $f'(z)$. Using the Cauchy integral formula for $n = 1$ with the function $f(z)$, we have the second integral also equal to $2\pi i\, f'(z_0)$.

An application of the Cauchy integral formulas yields a useful collection of inequalities known as the *Cauchy inequalities*.

Corollary 3.3   Hypotheses:
H1 $f(z)$ is analytic in a simply connected domain D.
H2 z_0 is a point of D and C is a circle centered at z_0 with radius $r > 0$ and contained in D.
H3   For all points z on C, $|f(z)| \leq M$.

Conclusion: For every non-negative integer n,

$$|f^{(n)}(z_0)| \leq n!\frac{M}{r^n}.$$

Proof: For each non-negative integer n, we use Theorem 3.4 to write

$$f^{(n)}(z_0) = \left(\frac{n!}{2\pi i}\right) \int_C \left[\frac{f(z)}{(z - z_0)^{n+1}}\right] dz,$$

where $C = \{z : |z - z_0| = r\}$. Writing $z - z_0 = r\,\exp(i\theta)$ for z on C, we have

$$f^{(n)}(z_0) = \frac{n!}{2\pi i} \int_0^{2\pi} \frac{f(z_0 + re^{i\theta})(ir\,e^{i\theta}\,d\theta)}{r^{n+1}\,\exp[i(n + 1)\theta]}$$

$$= \frac{n!}{2\pi i} \int_0^{2\pi} \frac{i\,f(z_0 + re^{i\theta})}{r^n\,\exp(in\theta)}\,d\theta$$

$$= \frac{n!}{2\pi r^n} \int_0^{2\pi} f(z_0 + re^{i\theta})e^{-in\theta}\,d\theta.$$

Then

$$|f^{(n)}(z_0)| \le \left(\frac{n!}{2\pi r^n}\right) \int_0^{2\pi} |f(z_0 + re^{i\theta})| \, d\theta$$

$$\le \frac{(2\pi n! \, M)}{(2\pi r^n)} = n! \, \frac{M}{r^n}.$$

Theorem 3.4 can tell us something more about analytic functions. For example, if $f(z)$ is analytic for all z, it will follow that $f(z)$ is either constant for all z or $|f(z)|$ can be made arbitrarily large; this will be the content of a statement known as *Liouville's Theorem*.

Corollary 3.4   Hypothesis: $f(z)$ is analytic for all z.

Conclusion: Either $f(z)$ is identically constant, or $|f(z)|$ is not bounded.

Proof: Suppose for some finite number $M > 0$ that $|f(z)| \le M$ for all z. Let z_0 be an arbitrary point; we shall use the previous corollary to show that $f'(z_0) = 0$. Then an application of problem 1.30 assures us that $f(z)$ is identically constant.

Let $C(r)$ be the circle $\{z : |z - z_0| = r\}$, where $r > 0$. Corollary 3.3 with $n = 1$ implies that $|f'(z_0)| \le M/r$. Since $f(z)$ — and hence $f'(z)$ — is analytic for all z, we can take the radius, r, of $C(r)$ to be arbitrarily large. This implies that $|f'(z_0)|$ can be made arbitrarily small. But since $f'(z_0)$ is a constant, we must conclude that $f'(z_0) = 0$, where z_0 was arbitrarily chosen.

Theorem 3.4 will also help us to prove a weak form of an important property of analytic functions known as the *maximum-modulus property*. If $f(z)$ is analytic in a simply-connected domain D with boundary C which is contained in some bounded region of the plane, then $|f(z)|$ is a real-valued continuous function of z on the closed and bounded subset of the plane consisting of D and C. Consequently we know $|f(z)|$ attains a maximum value at some point of $D \cup C$; the maximum-modulus theorem states that, unless $f(z)$ is identically constant throughout D, the point(s) where $|f(z)|$ attains this maximum value must lie on C — that is, must lie on the boundary of D. The proof of this statement is beyond our reach now, but in the following corollary we can prove a similar, slightly weaker, fact.

Corollary 3.5   Hypotheses:
H1 D is a disk $\{z : |z - z_0| < r\}$.

H2 $f(z)$ is analytic in D.

H3   For each z in D, $|f(z)| \leq |f(z_0)|$.

Conclusion: $f(z) \equiv f(z_0)$ for all z in D.

Proof: Choose any real number s, $0 < s < r$, and let $C(s)$ be the circle $\{z:|z - z_0| = s\}$. $C(s)$ lies in D, and by Theorem 3.4,

$$f(z_0) = (1/2\pi i) \int_{\check{C}(s)} \left[\frac{f(z)}{(z - z_0)} \right] dz$$

$$= (1/2\pi i) \int_0^{2\pi} \left[\frac{f(z_0 + se^{i\theta})}{se^{i\theta}} \right] (ise^{i\theta}\, d\theta)$$

$$= (1/2\pi) \int_0^{2\pi} f(z_0 + se^{i\theta})\, d\theta.$$

Then

$$|f(z_0)| \leq (1/2\pi) \int_0^{2\pi} |f(z_0 + se^{i\theta})|\, d\theta.$$

Since for all points $z_0 + se^{i\theta}$ we have $0 \leq |f(z_0 + se^{i\theta})| \leq |f(z_0)|$, and $|f(z_0 + se^{i\theta})|$ is surely continuous for $0 \leq \theta \leq 2\pi$, we must have $|f(z_0 + se^{i\theta})| = |f(z_0)|$ for all θ, $0 \leq \theta \leq 2\pi$. But any point of D lies on a circle $C(s)$ about z_0 for some s, $0 \leq s < r$, so we see that $|f(z)| = |f(z_0)|$ for all z in D. This is enough to tell us that $f(z) = f(z_0)$ for all z in D because of problem 1.27.

One of the consequences of the maximum-modulus theorem for a function $f(z)$, analytic in a simply-connected domain D with boundary C is that it tells us we can find the maximum of $|f(z)|$ in $D \cup C$ by finding the maximum of $|f(z)|$ on C alone.

Example 3.11   Let D be the rectangle $\{(x,y):0 < x < 2\pi, 0 < y < 1\}$ and C be its boundary. The function $f(z) = \sin z$ is analytic at each point of $D \cup C$, and $|f(z)|^2 = \sin^2 x + \sinh^2 y$. The reader can verify that $|f(z)|^2$ has a maximum value of $1 + \sinh^2 1$ on $D \cup C$ which is attained at the points $\pi/2 + i$, $3\pi/2 + i$.

To conclude this section we state a theorem which gives us a criterion for deciding when a continuous function is actually analytic. This theorem is almost a converse to the Cauchy Integral Theorem.

THEOREM 3.5   Morera's Theorem
Hypotheses:
H1 $f(z)$ is continuous in a simply-connected domain D.

H2   For any closed path C in D, $\displaystyle\int_C f(z)\, dz = 0$.

Conclusion: $f(z)$ is analytic in D.

Proof: H2 is equivalent to saying that for every pair of points, a and b, in D, $\displaystyle\int_a^b f(z)dz$ is independent of path in D. Since $f(z)$ is continuous in D, C1 of Theorem 3.1 tells us $f(z)$ is the derivative of a function analytic in D. That $f(z)$ is analytic in D now follows from Theorem 3.4.

PROBLEMS

3.17 Evaluate each of the following integrals.

(a) $\displaystyle\int_C [\cos z/(z - \pi/2)^2]\, dz$, where C is the circle $\{z : |z| = 2\}$.

(b) $\displaystyle\int_C [e^z/(z - 1)(z - 2)]\, dz$, where C is the circle $\{z : |z| = 3\}$.

(c) $\displaystyle\int_C [z^n/(z - a)^{n+1}]\, dz$, where C is any closed path enclosing $z = a$.

(d) $\displaystyle\int_C [(z^2 + 1)/(z - 2)^3]\, dz$, where C is any closed path not passing through $z = 2$.

(e) $\displaystyle\int_C z^{m-n-1}\, dz$, if C is the circle $\{z : |z| = 1\}$, and m, n are integers.

3.18 Suppose $f(z)$ is analytic in a domain containing a closed path C and its interior. If z_1 and z_2 are any points inside C, show that

$$f(z_1) = f(z_2) + \frac{(z_1 - z_2)}{2\pi i} \int_C \frac{f(z)}{(z - z_1)(z - z_2)} \, dz.$$

3.19 If $f(z)$ is analytic for all z and $|f(z)| \leq M$ for all z and some finite number M, use the previous problem to show that $f(z) = f(0)$ for all z. (*Hint:* For any z_1, z_2, take C to be a circle of radius $R >$ max $\{|z_1|, |z_2|\}$ and estimate $|f(z_1) - f(z_2)|$ as R increases without bound.) This gives an alternative proof of Liouville's Theorem.

3.20 Use Liouville's Theorem to prove the Fundamental Theorem of Algebra: if $p(z) = a_0 + a_1 z + \cdots + a_n z^n$ is any polynomial of degree $n \geq 1$, then there exists at least one point z_0 such that $p(z_0) = 0$.

3.21 If $f(z)$ is analytic for $|z| < 3$ and $|f(z)| \leq 5$ for $|z| = 2$, what is the largest possible value for $|f^{(2)}(\tfrac{1}{2})|$?

3.22 If W is a closed bounded set in the complex plane, at what point(s) of W will e^z have its maximum modulus?

3.23 If $u(x,y)$ is a harmonic function in $D = \{(x,y):(x - x_0)^2 + (y - y_0)^2 < r^2\}$ and $0 \leq u(x,y) \leq u(x_0,y_0)$ for all points (x,y) of D, prove that $u(x,y) \equiv u(x_0,y_0)$ in D. That is, harmonic functions also have the kind of behavior described in Corollary 3.5 for analytic functions. (*Hint:* Use Theorem 1.8 to construct a function $f(z)$ analytic in D with $u(x,y) = \text{Re } f(z)$. Then examine the function $\exp[f(z)]$.)

3.24 Suppose $f(z)$ is analytic for z in $D = \{z:|z| \leq 1\}$ while $|f(z)| \geq |f(0)|$ for all z in D.

(a) Find an example to show that it is not necessarily true that $f(z) = f(0)$ throughout D.

(b) What additional hypothesis on $f(z)$ will allow you to prove that $f(z) = f(0)$ everywhere in D?

3.25 Use the preceding problem to help you formulate a minimum-modulus principle for analytic functions.

3.26 If $f(z) = z^2 + 1$ for $W = \{z:|z| \leq 1\}$, find the maximum and minimum values for $|f(z)|$ in W and the points in W where these extreme values are attained.

4

Infinite Series

We have discussed what it means for a function of a complex variable to be analytic and have seen how analytic functions arise naturally from integrating functions along paths. There is another approach to analytic functions which is of equal importance with those we have studied. This approach is through power series, and with it we shall be able to see even better what it means for a function to be analytic.

Even more, we shall be in a position to examine a function which is analytic except at isolated points, and, in a sense, to decide how close to being analytic the function is at those points. For example, the function $f(z) = 1/z$ is analytic for all z except $z = 0$, and the same is true for $g(z) = 1/z^k$, where k is a positive integer. Surely a variety of functions are analytic for $0 < |z| < 1$ and not analytic at $z = 0$, and we want to see as well as we can what properties all such functions have in common.

Specifically, in this chapter we shall see how power series lead us to analytic functions. As a means of discussing how a function might be analytic for $0 < |z - a| < r$ and not analytic at $z = a$, we shall study an additional type of power series known as a Laurent's series.

Section 4.1   Power Series

The entrance fee for this section, and the entire chapter, is a knowledge of tests for convergence of infinite series of real constants and the facts about functions of a real variable represented by power series. For those with a grasp of this background the progression of results will be very familiar and orderly.

If $\{a_n\}_{n=0}^{\infty} = \{a_0, a_1, \ldots, a_n, \ldots\}$ is a sequence of complex numbers, the infinite series $\sum_{n=0}^{\infty} a_n$ is said to *converge to S* if:

(1) the sequence $\left\{ S_n = \sum_{j=0}^{n} a_j \right\}_{n=0}^{\infty}$ has limit S; or equivalently,

(2) for every number $\epsilon > 0$ there exists an integer $N(\epsilon) > 0$ such that $|S - (a_0 + a_1 + \cdots + a_n)| < \epsilon$ whenever $n > N(\epsilon)$.

If the series $\sum_{n=0}^{\infty} a_n$ does not converge to some number, we say that it *diverges*.

We say that $\sum_{n=0}^{\infty} a_n$ *converges absolutely* if the series $\sum_{n=0}^{\infty} |a_n|$ converges. All the familiar tests for convergence of series of nonnegative real constants can be used as tests for absolute convergence of series of complex constants.

Our exclusive interest in this section is the collection of functions defined by power series.

Definition   If $\{a_n\}_{n=0}^{\infty}$ is a sequence of complex numbers, we call $\sum_{n=0}^{\infty} a_n(z - a)^n$ a *(complex) power series about $z = a$*.

We deal explicitly with only the case where $a = 0$, but the reader will realize each statement we make has an obvious analogue for cases where $a \neq 0$.

To say the power series $\sum_{n=0}^{\infty} a_n z^n$ is convergent at $z = z_0$ of course means that the series of constants $\sum_{n=0}^{\infty} a_n z_0^n$ is convergent. The reader can verify for himself that if $\sum_{n=0}^{\infty} a_n z^n$ is convergent at $z = z_0$, then it is absolutely convergent for all z, $0 \leq |z| < |z_0|$. Thus, if $\sum_{n=0}^{\infty} a_n z^n$ is convergent for any nonzero z, it is absolutely convergent at every point of an open disk centered at $z = 0$. (The reader might stop here to write out a comparable statement for a power series $\sum_{n=0}^{\infty} a_n(z - a)^n$ at $z = a$.) We naturally seek the largest possible "disk of convergence" to be associated with the power series, and the candidate for the radius of this largest possible disk, as given in the following definition, should be familiar.

Definition   Let $\sum_{n=0}^{\infty} a_n z^n$ be a power series, and let

$$S = \{r \geq 0: \sum_{n=0}^{\infty} |a_n| r^n \text{ is convergent}\}.$$

If S is bounded above, we define the *radius of convergence*, ρ, of the power series to be the least upper bound of the set S. If S is not bounded above, we say $\sum_{n=0}^{\infty} a_n z^n$ has infinite radius of convergence and write $\rho = \infty$. In the first case we call $\{z: |z| < \rho\}$ the *disk of convergence* for $\sum_{n=0}^{\infty} a_n z^n$, while in the second case we say $\sum_{n=0}^{\infty} a_n z^n$ is *convergent everywhere*.

Example 4.1   The power series $\sum_{n=0}^{\infty} z^n/n!$ (with $0! = 1$) has infinite radius of convergence, since the ratio test will show that $\sum_{n=0}^{\infty} r^n/n!$ is convergent for any $r \geq 0$ and this shows that the set S in the definition is $[0, \infty)$.

Definition   Let $\sum_{n=0}^{\infty} a_n z^n$ be a power series and D be a set of complex numbers. We say $\sum_{n=0}^{\infty} a_n z^n$ is *uniformly convergent on D* if there is a function $s(z)$ defined on D and for any positive number ϵ there exists a positive integer $N(\epsilon, D)$ such that:
  (1) $|s(z) - \sum_{n=0}^{k} a_n z^n| < \epsilon$ whenever $k \geq N$ and $z \in D$; or equivalently,
  (2) $|\sum_{n=k+1}^{\infty} a_n z^n| < \epsilon$ whenever $k \geq N$ and $z \in D$.

If $\sum_{n=0}^{\infty} a_n z^n$ is uniformly convergent on D, it is surely convergent at each point of D and has sum $s(z)$, so we frequently say $\sum_{n=0}^{\infty} a_n z^n$ converges uniformly to $s(z)$ on D.

Now let $\sum_{n=0}^{\infty} a_n z^n$ be a power series with radius of convergence ρ. We deal explicitly with the case $0 < \rho < \infty$, and the reader will have no trouble with the additional cases where $\rho = 0$ or $\rho = \infty$. In the open disk $\{z: |z| < \rho\}$ the power series is absolutely convergent and so represents some function of z; let us write

$$P(z) = \sum_{n=0}^{\infty} a_n z^n \text{ for } |z| < \rho.$$

The following theorems lead to our main conclusion that $P(z)$ is analytic for $|z| < \rho$, as well as the fact that in $\{z: |z| < \rho\}$ $P'(z)$ is given by the series $\sum_{n=1}^{\infty} n a_n z^{n-1}$ with radius of convergence ρ.

THEOREM 4.1   Hypothesis: $P(z) = \sum_{n=0}^{\infty} a_n z^n$ is a power series with radius of convergence ρ.

Conclusions:
C1   For each z, $|z| < \rho$, $\sum_{n=0}^{\infty} a_n z^n$ is absolutely convergent.
C2   For each z, $|z| > \rho$, $\sum_{n=0}^{\infty} a_n z^n$ is divergent.
C3   If D is a compact subset of $\{z : |z| < \rho\}$, $\sum_{n=0}^{\infty} a_n z^n$ is uniformly convergent on D.
C4 $P(z)$ is continuous in $\{z : |z| < \rho\}$.

Proof: We have already remarked that the statement C1 holds. Were $\sum_{n=0}^{\infty} a_n z_0^n$ to be convergent for any z_0 with $|z_0| > \rho$, $\sum_{n=0}^{\infty} a_n z^n$ would be absolutely convergent for all z such that $|z| < |z_0|$. In particular, $\sum_{n=0}^{\infty} a_n z_1^n$ would be absolutely convergent for $z_1 = (\rho + |z_0|)/2 > \rho$, and this contradicts our definition of ρ. Thus C2 is verified.

If D is a compact subset of $\{z : |z| < \rho\}$, we can find a positive number γ, $0 < \gamma < \rho$, such that $D \subset \{z : |z| < \gamma\} \subset \{z : |z| < \rho\}$. (See Fig. 4.1.) (The proof of this statement takes a technical argument not in the main line of the present proof, so we do not give any details.) Since $0 < \gamma < \rho$, $\sum_{n=0}^{\infty} |a_n| \gamma^n$ is convergent; thus, given any number $\epsilon > 0$, there exists a positive integer N such that whenever $k \geq N$,

$$\sum_{n=k+1}^{\infty} |a_n| \gamma^n < \epsilon.$$

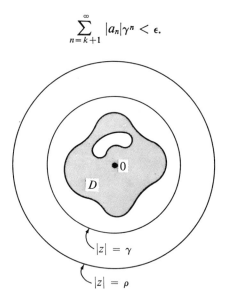

Figure 4.1

Now for all $z \in D$ and any integer $k \geq N$, $|z| \leq \gamma$, and

$$\left| P(z) - \sum_{n=0}^{k} a_n z^n \right| = \left| \sum_{n=k+1}^{\infty} a_n z^n \right| \leq \sum_{n=k+1}^{\infty} |a_n||z|^n$$

$$\leq \sum_{n=k+1}^{\infty} |a_n| \gamma^n < \epsilon.$$

Thus, $\sum_{n=0}^{\infty} a_n z^n$ is uniformly convergent on D (in fact on $\{z : |z| \leq \gamma\}$.)

Now each term of the power series $P(z)$ is continuous in $\{z : |z| < \rho\}$, and the series for $P(z)$ is uniformly convergent in any compact subset of $\{z : |z| < \rho\}$. To establish C4 we must show that these facts imply that $P(z)$ is continuous at each point of $\{z : |z| < \rho\}$.

Let z_0 be any point with $|z_0| < \rho$. We choose as a compact subset of $\{z : |z| < \rho\}$ the set $D = \{z : |z - z_0| \leq (\rho - |z_0|)/2\}$; D contains z_0 and is contained in $\{z : |z| < \rho\}$ (see Fig. 4.2). Given any number $\epsilon > 0$, we can find a positive integer $N(\epsilon, D)$ such that for all z in D

$$\left| P(z) - \sum_{n=0}^{N} a_n z^n \right| < \frac{\epsilon}{3}.$$

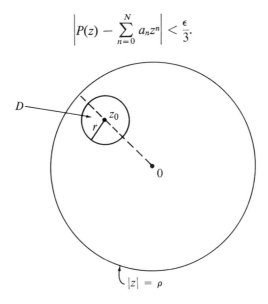

Figure 4.2

In particular, for $z = z_0$, $|P(z_0) - \sum_{n=0}^{N} a_n z_0^n| < \epsilon/3$.

Notice that the polynomial $\sum_{n=0}^{N} a_n z^n$ is continuous throughout D, and we can find a positive number $\delta < (\rho - |z_0|)/2$ so that $|\sum_{n=0}^{N} a_n z^n - \sum_{n=0}^{N} a_n z_0^n| < \epsilon/3$ whenever $|z - z_0| < \delta$. Then for all z in D with $|z - z_0| < \delta$ we have

$$\left| P(z) - P(z_0) \right| \leq \left| P(z) - \sum_{n=0}^{N} a_n z^n \right| + \left| \sum_{n=0}^{N} a_n z^n - \sum_{n=0}^{N} a_n z_0^n \right|$$

$$+ \left| \sum_{n=0}^{N} a_n z_0 - P(z_0) \right|$$

$$< \frac{\epsilon}{3} + \frac{\epsilon}{3} + \frac{\epsilon}{3} = \epsilon.$$

This shows $P(z)$ is continuous at $z = z_0$, an arbitrary point of $\{z : |z| < \rho\}$.

Now that we know $P(z)$ is continuous in $\{z : |z| < \rho\}$, we can be sure $\int_C P(z)\, dz$ exists, where C is any path lying in $\{z : |z| < \rho\}$.

THEOREM 4.2 Hypothesis: $P(z) = \sum_{n=0}^{\infty} a_n z^n$ is a power series with radius of convergence ρ.

Conclusion: For any path C lying in $\{z : |z| < \rho\}$,

$$\int_C P(z)\, dz = \int_C \left(\sum_{n=0}^{\infty} a_n z^n \right) dz = \sum_{n=0}^{\infty} \left[\int_C a_n z^n\, dz \right].$$

Outline of Proof: As a compact subset D of $\{z : |z| < \rho\}$ we take the points of the path C. If L is the length of C and ϵ is an arbitrary positive number, there exists a positive integer N, depending on ϵ and D, such that

$$\left| P(z) - \sum_{n=0}^{k} a_n z^n \right| < \frac{\epsilon}{L}$$

whenever $k \geq N$ and $z \in D$. Now for each $k \geq N$,

$$\left| \int_C P(z)\, dz - \sum_{n=0}^{k} \left[\int_C a_n z^n\, dz \right] \right| = \left| \int_C P(z)\, dz - \int_C \left[\sum_{n=0}^{k} a_n z^n \right] dz \right|$$

$$= \left| \int_C \left[P(z) - \sum_{n=0}^{k} a_n z^n \right] dz \right|$$

$$\leq \int_C \left| P(z) - \sum_{n=0}^{k} a_n z^n \right| |dz| < \left(\frac{\epsilon}{L} \right) L = \epsilon.$$

This proves the theorem.

It is worthwhile to point out two consequences of this theorem.

    (1) We can integrate a power series term-by-term along any path in the region of convergence.

    (2) Since each term of the power series is of the form $a_n z^n$, for any path C in the region of convergence $\int_C a_n z^n \, dz$ is very easy to evaluate. In fact, for $P(z) = \sum_{n=0}^{\infty} a_n z^n$, it may be easier to evaluate $\int_C P(z) \, dz$ by evaluating $\sum_{n=0}^{\infty} \left[\int_C a_n z^n \, dz \right]$.

The previous two theorems leave us in position to obtain the conclusion that $P(z) = \sum_{n=0}^{\infty} a_n z^n$ is analytic in its region of convergence.

THEOREM 4.3 Hypothesis: $P(z) = \sum_{n=0}^{\infty} a_n z^n$ is a power series with radius of convergence ρ.

Conclusions:

C1 $P(z)$ is analytic in $\{z : |z| < \rho\}$.

C2 The power series $\sum_{n=1}^{\infty} n a_n z^{n-1}$ (formed from the term-by-term differentiation of $P(z)$) has radius of convergence ρ and is uniformly convergent to $P'(z)$ on each compact subset of $\{z : |z| < \rho\}$. That is, for $|z| < \rho$,

$$P'(z) = \sum_{n=1}^{\infty} n a_n z^{n-1}.$$

Proof: Let C be any closed path in $\{z : |z| < \rho\}$. By the previous theorem $\int_C P(z) \, dz = \sum_{n=0}^{\infty} \left[\int_C a_n z^n \, dz \right] = \sum_{n=0}^{\infty} \left[a_n \int_C z^n \, dz \right] = 0$. Now $P(z)$ is continuous in $\{z : |z| < \rho\}$ by Theorem 4.1, so from Morera's Theorem we can see $P(z)$ is analytic in $\{z : |z| < \rho\}$.

In proving C2 we have the Cauchy integral formula available to help us represent $P'(z)$ in terms of a line integral involving $P(z)$.

Let z_0 be any point of $\{z : |z| < \rho\}$ and let C be the circle $\{z : |z - z_0| = (\rho - |z_0|)^2/2\}$. Then we may write $P'(z_0) = (1/2\pi i) \int_C [P(z)/(z - z_0)^2] \, dz$, and for each positive integer n

$$n a_n z_0^{n-1} = (1/2\pi i) \int_C \left[\frac{a_n z^n}{(z - z_0)^2} \right] dz.$$

The points of C comprise a compact subset of $\{z : |z| < \rho\}$, and given any number $\epsilon > 0$ there exists a positive integer N, depending on ϵ and C, such that $|P(z) - \sum_{n=0}^{k} a_n z^n| < \epsilon$ whenever $k \geq N$ and $z \in C$. Now for any $k \geq N$,

$$\left| P'(z_0) - \sum_{n=1}^{k} n a_n z^{n-1} \right| = \left| (1/2\pi i) \int_C \left[\frac{P(z)}{(z - z_0)^2} \right] dz - (1/2\pi i) \int_C \frac{\sum_{n=0}^{k} a_n z^n}{(z - z_0)^2} \, dz \right|$$

$$= \left| (1/2\pi i) \int_C \left\{ \frac{P(z) - \sum_{n=0}^{k} a_n z^n}{(z - z_0)^2} \right\} dz \right|$$

$$\leq (1/2\pi) \int_C \left\{ \frac{|P(z) - \sum_{n=0}^{k} a_n z^n|}{|z - z_0|^2} \right\} |dz|$$

$$< (1/2\pi) \left[\frac{2\epsilon}{(\rho - |z_0|)^2} \right] \int_C |dz| = \frac{2\epsilon}{\rho - |z_0|^2}.$$

Thus for each z_0 with $|z_0| < \rho$ we see that

$$P'(z_0) = \sum_{n=1}^{\infty} n a_n z_0^{n-1}.$$

Theorems 4.1, 4.2, and 4.3 may be applied in turn to the power series $P'(z) = \sum_{n=1}^{\infty} n a_n z^{n-1}$ in the disk of convergence $\{z : |z| < \rho\}$ to find power series representations for all the higher order derivatives of $P(z)$.

Before we look at some examples there are two points to be made:

(1) For a power series $\sum_{n=0}^{\infty} a_n z^n$ with radius of convergence ρ the three theorems above have made no comment at all about the convergence properties for the series on the circle of convergence $\{z : |z| = \rho\}$ separating the disk of convergence from the region where the series is divergent. Examples below will help to explain this seeming omission.

(2) More explicit means of deciding on the value of ρ would be helpful — and are surely available in the case of real power series.

To settle the second point we state together two results without proof. The results and their proofs differ very little from those found in a treatment of real power series.

THEOREM 4.4 Hypothesis: Let $\sum_{n=0}^{\infty} a_n z^n$ be a power series with radius of convergence ρ, $0 \leq \rho \leq +\infty$.

Conclusions:
C1 If the sequence $\{|a_{n+1}|/|a_n|\}_{n=1}^{\infty}$ converges, its limit is $1/\rho$. That is,

$$\rho = 1/\lim_{n \to \infty} \left(\frac{|a_{n+1}|}{|a_n|} \right)$$

if this limit exists.
C2 If the sequence $\{|a_n|^{1/n}\}_{n=1}^{\infty}$ converges, its limit is $1/\rho$. That is,

$$\rho = 1/\lim_{n \to \infty} (|a_n|^{1/n})$$

if this limit exists.

In the following three examples we examine three power series with radius of convergence 1 whose behavior on the circle $\{z : |z| = 1\}$ shows some variety.

Example 4.2 Let $P_1(z) = \sum_{n=0}^{\infty} z^n$. For each positive integer n, $a_n = 1$, and obviously $\rho = 1$.
On the circle $\{z : |z| = 1\}$ the series is divergent at every point, because for no point z_0 with $|z_0| = 1$ can it be said that $\lim_{n \to \infty} z_0^n = 0$.

Example 4.3 Let $P_2(z) = \sum_{n=1}^{\infty} z^n/n$. For $a_n = 1/n$, $\lim_{n \to \infty}(|a_{n+1}|/|a_n|) = 1 = 1/\rho$, and $\rho = 1$.
The series for $P_2(z)$ is convergent for $z = -1$, divergent for $z = 1$. The series is not absolutely convergent at any point of $\{z : |z| = 1\}$.

Example 4.4 Let $P_3(z) = \sum_{n=1}^{\infty} z^n/n^2$. For $a_n = 1/n^2$, $\rho = 1/\lim_{n \to \infty}(|a_{n+1}|/|a_n|) = 1$, and the power series is absolutely convergent at every point of $\{z : |z| = 1\}$.

Example 4.5 In Example 4.2 we saw that $P_1(z) = \sum_{n=0}^{\infty} z^n$ represents an analytic function in the disk $\{z : |z| < 1\}$. For each nonnegative integer N, $\sum_{n=0}^{N} z^n = (1 - z^{N+1})/(1 - z)$, so for each z, $|z| < 1$,

$$P_1(z) = \lim_{N \to \infty} \frac{(1 - z^{N+1})}{(1 - z)}$$

$$= \frac{1}{(1 - z)} - \lim_{N \to \infty} \frac{z^{N+1}}{(1 - z)} = \frac{1}{(1 - z)}.$$

That is, $\sum_{n=0}^{\infty} z^n = 1/(1 - z)$ for $|z| < 1$. If we apply Theorems 4.3 and 4.2 to this result, we may write

$$\sum_{n=1}^{\infty} n z^{n-1} = \frac{1}{(1 - z)^2} \text{ for } |z| < 1$$

and

$$\sum_{n=0}^{\infty} \frac{z^{n+1}}{(n + 1)} = -\text{Log}(1 - z) \text{ for } |z| < 1.$$

To indicate one way in which our discussion of power series can be brought to bear on infinite series of functions which are not in the form of power series, we want to prove and illustrate a theorem which can be extensively generalized.

THEOREM 4.5 Hypotheses: For a simply connected domain D and some positive $M, f(z)$ is analytic in D and $|f(z)| < M$ at all points of D.

Conclusion: The infinite series $\sum_{n=0}^{\infty} [f(z)/M]^n$ is an analytic function in D, and at each point of D

$$\sum_{n=0}^{\infty} \left[\frac{f(z)}{M} \right]^n = \frac{M}{[M - f(z)]}.$$

Proof: If $w = f(z)/M$ for each $z \in D$, then $|w| < 1$. Define $F(w) = \sum_{n=0}^{\infty} w^n$ for each w, $|w| < 1$. From Example 4.5 we know that $F(w) = 1/(1 - w)$ in $\{w:|w| < 1\}$.

Since the range of $w = f(z)/M$ is contained in $\{w:|w| < 1\}$ and $w = f(z)/M$ is analytic in D, by Theorem 1.6 $G(z) = F[f(z)/M]$ is analytic in D. That is, $G(z) = \sum_{n=0}^{\infty} [f(z)/M]^n$ is analytic in D and for each z in D,

$$G(z) = F\left[\frac{f(z)}{M} \right] = \frac{1}{1 - \dfrac{f(z)}{M}} = \frac{M}{M - f(z)}.$$

Example 4.6 If $f(z) = \exp(z - 1)$ and $D = \{z:|z| < 1\}$, $|f(z)| < 1$ in D, and

$$\sum_{n=0}^{\infty} \exp n(z - 1) = \frac{1}{1 - \exp(z - 1)} = \frac{e}{e - e^z}$$

is analytic in D.

Example 4.7 The function $f(z) = 1/z$ is analytic for $|z| > 1$, and $|f(z)| < 1$ whenever $|z| > 1$. Then Theorem 4.5 implies that

$$\sum_{n=0}^{\infty} \frac{1}{z^n} = \frac{z}{(z-1)} \quad \text{for all } z, \ |z| > 1.$$

(In our narrow use of the phrase "power series" the series $\sum_{n=0}^{\infty} 1/z^n$ is not a power series, although it is a series of (negative) powers of z. This type of series will be treated more fully later.)

Example 4.8   For a fixed positive number M, let D be the strip $\{z: -M < \text{Im } z < M\}$ in the complex plane. For any $z \in D$, $|\sin z| \le [\exp(-\text{Im } z) + \exp(\text{Im } z)]/2 < e^M$, and $\sin z$ is analytic in D. Then

$$\sum_{n=0}^{\infty} \left[\frac{\sin z}{e^M} \right]^n = \frac{e^M}{(e^M - \sin z)}$$

is analytic in D.

PROBLEMS

4.1   Find the radius of convergence and disk of convergence of each of the following power series.

(a) $\sum_{n=0}^{\infty} z^n/n!$

(b) $\sum_{n=0}^{\infty} [(-1)^n z^n/(2n+1)!]$

(c) $\sum_{n=0}^{\infty} (n+1)(z-2)^n$

(d) $\sum_{n=0}^{\infty} n!(z-1)^n$

4.2   None of the infinite series below is what we have called a power series, yet each can be rewritten in power series form. By so rewriting and finding the appropriate disk of convergence, determine a region in which each series represents an analytic function.

(a) $\sum_{n=1}^{\infty} n(2z+3)^n/3^n$

(b) $\sum_{n=0}^{\infty} n(z^2 + 2z + 1)^n/2^n$

(c) $\sum_{n=1}^{\infty} n(3z+2)^n$

4.3   Suppose that $\sum_{n=0}^{\infty} a_n z^n$ has radius of convergence ρ_1 and $\sum_{n=0}^{\infty} b_n z^n$ has radius of convergence ρ_2. If α and β are any constants, show that in the disk $\{z: |z| < \min(\rho_1, \rho_2)\}$

$$\sum_{n=0}^{\infty} (\alpha a_n + \beta b_n) z^n = \alpha \sum_{n=0}^{\infty} a_n z^n + \beta \sum_{n=0}^{\infty} b_n z^n.$$

4.4   For each of the following series find the disk of convergence within
       which the series represents an analytic function. Use Theorems 4.2
       and 4.3, together with the fact that $\sum_{n=0}^{\infty} z^n = 1/(1 - z)$ for
       $|z| < 1$, to determine the function explicitly.

(a) $\sum_{n=0}^{\infty} z^{n+1}/(n + 1)$

(b) $\sum_{n=1}^{\infty} n^2 z^n$

(c) $\sum_{n=1}^{\infty} z^n/n^2$

(d) $\sum_{n=1}^{\infty} n(3z + 2)^n$ (*Hint:* Let $w = 3z + 2$.)

4.5   Show that the series $\sum_{u=0}^{\infty} 1/(3z + 5)^n$ represents the analytic
       function $(3z + 5)/(3z + 4)$ in the region $\{z: |z + \frac{5}{3}| > \frac{1}{3}\}$.

4.6   Determine the region of convergence for the series

$$\sum_{n=0}^{\infty} \frac{\{\exp[2\pi i(n + 1)z]\}}{(n + 1)}.$$

What analytic function does it represent in this region?

Section 4.2   Taylor's Series

If $f(z)$ is analytic at $z = a$, then $f(z)$ has derivatives of all orders at
$z = a$, and it is natural for us to think of Taylor's series as a means of
representing analytic functions. The existence of derivatives of all orders
for a real-valued function at one point on the real axis is not in itself
enough to guarantee the existence of a Taylor's series representation of
the function in a neighborhood of that point. A well-known example
is the function

$$f(x) = \begin{cases} \exp\left(\dfrac{-1}{x^2}\right), & x \neq 0 \\ 0, & x = 0 \end{cases}$$

which has derivatives of all orders at $x = 0$, all of which equal zero.
    We shall see that such unpleasant situations will not arise for us in
considering analytic functions in the complex plane, and we shall find
that if $f(z)$ is analytic at $z = a$, $f(z)$ does have a Taylor's series representation valid in some open disk centered at $z = a$. Suppose, in particular, that $f(z)$ is written in the form of a power series, $\sum_{n=0}^{\infty} a_n(z - a)^n$,
having a radius of convergence $\rho > 0$, so that $f(z)$ is an analytic function

in the disk $\{z:|z - a| < \rho\}$. This analytic function will have a Taylor's series expansion valid in some disk centered at $z = a$. If the Taylor's series were to differ in any way from the power series $\sum_{n=0}^{\infty} a_n(z - a)^n$ at points near $z = a$, the job of gleaning information about the analytic functions from their power series representations could be quite complicated. We shall see that in the vicinity of a given point an analytic function has only one power series expansion, and that is the Taylor's series.

THEOREM 4.6   Hypothesis: For a fixed point $z = a$ and a positive number r, $f(z)$ is analytic in $\{z:|z - a| < r\}$.

Conclusion: For any point z with $|z - a| < r$,

$$f(z) = f(a) + \sum_{n=1}^{\infty} \frac{f^{(n)}(a)}{n!}(z - a)^n.$$

(We call this representation of $f(z)$ in $\{z:|z - a| < r\}$ the *Taylor's series for $f(z)$ about $z = a$*. As in the real case, when $a = 0$, the resulting representation may also be called the *Maclaurin's series for $f(z)$*.)

Proof: Let z_0 be any point in $\{z:|z - a| < r\}$ and let $s = |z_0 - a|$. For any t, $s < t < r$, let C be the circle $\{z:|z - a| = t\}$. We can write

$$f(z_0) = (1/2\pi i) \int_C \frac{f(z)}{(z - z_0)}\, dz.$$

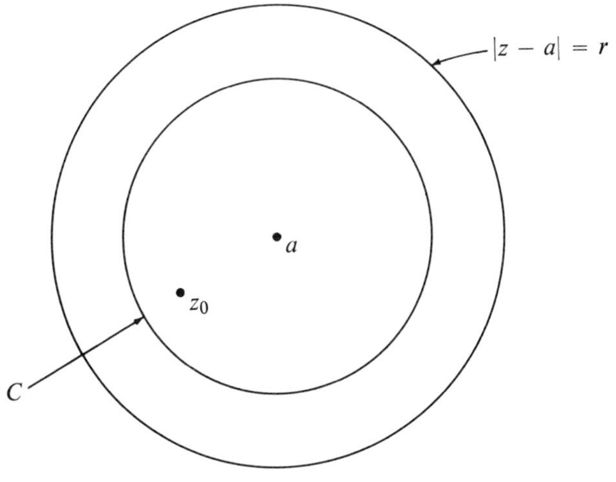

Figure 4.3

We first try to rewrite $1/(z - z_0)$ in powers of $(z_0 - a)$ by noting that

$$z - z_0 = (z - a) - (z_0 - a)$$

$$= (z - a)\left[1 - \left(\frac{z_0 - a}{z - a}\right)\right],$$

where
$$\left|\frac{z_0 - a}{z - a}\right| = \frac{s}{t} < 1.$$

Now for each positive integer k,

$$\frac{1}{z - z_0} = \frac{1}{(z - a)\left[1 - \left(\frac{z_0 - a}{z - a}\right)\right]}$$

$$= \left(\frac{1}{z - a}\right)\left\{1 + \left(\frac{z_0 - a}{z - a}\right) + \left(\frac{z_0 - a}{z - a}\right)^2 + \cdots + \left(\frac{z_0 - a}{z - a}\right)^{k-1}\right.$$

$$\left. + \left(\frac{z_0 - a}{z - a}\right)^k\left[\frac{1}{1 - \left(\frac{z_0 - a}{z - a}\right)}\right]\right\}.$$

Consequently

$$f(z_0) = \frac{1}{2\pi i}\int_C \frac{f(z)}{z - z_0}\,dz$$

$$= \frac{1}{2\pi i}\int_C \frac{f(z)}{z - a}\,dz + \frac{1}{2\pi i}(z_0 - a)\int_C \frac{f(z)}{(z - a)^2}\,dz$$

$$+ \frac{1}{2\pi i}(z_0 - a)^2\int_C \frac{f(z)}{(z - a)^3}\,dz + \cdots$$

$$+ \frac{1}{2\pi i}(z_0 - a)^{k-1}\int_C \frac{f(z)}{(z - a)^k}\,dz$$

$$+ \frac{1}{2\pi i}(z_0 - a)^k\int_C \frac{f(z)}{(z - a)^k\left[1 - \left(\frac{z_0 - a}{z - a}\right)\right]}\,dz$$

$$= f(a) + f'(a)(z_0 - a) + \frac{f''(a)}{2!}(z_0 - a)^2 + \cdots$$

$$+ \frac{f^{(k-1)}(a)}{(k-1)!}(z_0 - a)^{k-1}$$

$$+ \frac{1}{2\pi i}(z_0 - a)^k \int_C \frac{f(z)}{(z-a)^k \left[1 - \left(\dfrac{z_0 - a}{z - a}\right)\right]} \, dz.$$

If $f^{(0)}(a) = f(a)$, and $0! = 1$,

$$\left| f(z_0) - \sum_{n=0}^{k} \frac{f^{(n)}(a)}{n!}(z_0 - a)^n \right|$$

$$= \left| \frac{1}{2\pi i}(z_0 - a)^k \int_C \frac{f(z)}{(z-a)^k \left[1 - \left(\dfrac{z_0 - a}{z - a}\right)\right]} \, dz \right|.$$

On C, $f(z)$ is continuous, so for some number $M > 0$, $|f(z)| \le M$ whenever $z \in C$. Also for $z \in C$, $|z - a| = t$, and $|1 - [(z_0 - a)/(z - a)]| \ge 1 - s/t > 0$. Thus

$$\left| \frac{1}{2\pi i}(z_0 - a)^k \int_C \frac{f(z)}{(z-a)^k \left[1 - \left(\dfrac{z_0 - a}{z - a}\right)\right]} \, dz \right|$$

$$\le \frac{1}{2\pi}|z_0 - a|^k \int_C \frac{|f(z)|\,|dz|}{|z - a|^k \left|1 - \left(\dfrac{z_0 - a}{z - a}\right)\right|}$$

$$\le \frac{1}{2\pi} \frac{s^k M}{t^k \left(1 - \dfrac{s}{t}\right)}(2\pi t)$$

$$= \left(\frac{t^2 M}{t - s}\right)\left(\frac{s}{t}\right)^k.$$

Given any number $\epsilon > 0$, since $0 < s/t < 1$, there exists an integer K, depending on ϵ, M, t, s, such that whenever $k > K$,

$$\left(\frac{s}{t}\right)^k < \frac{\epsilon(t - s)}{t^2 M} \quad \text{and} \quad \left(\frac{t^2 M}{t - s}\right)\left(\frac{s}{t}\right)^k < \epsilon.$$

That is, for $k > K$,

$$\left| f(z_0) - \sum_{n=0}^{k} \frac{f^{(n)}(a)}{n!}(z_0 - a)^n \right| < \epsilon,$$

or equivalently,

$$f(z_0) = \sum_{n=0}^{\infty} \frac{f^{(n)}(a)}{n!} (z_0 - a)^n \quad \text{whenever} \quad |z_0 - a| < r.$$

Remarks:
  (1)   Under the hypothesis that $f(z)$ is analytic for $|z - a| < r$, the
        series expansion above is valid throughout $|z - a| < r$. This
        hypothesis is the only restraint on r, so r can be as large as the
        distance from $z = a$ to the nearest point where $f(z)$ is not
        analytic. (If $f(z)$ is analytic everywhere, we can take $r = \infty$.)
        This shows that the Taylor's series expansion for $f(z)$ about
        $z = a$ is valid at all points inside a circle centered at $z = a$
        passing through the point nearest to $z = a$ where $f(z)$ is
        not analytic.
  (2)   The properties we have already derived for power series imply
        that within the circle just described the Taylor's series for
        $f(z)$ about $z = a$:
        (a)   is absolutely convergent at each point;
        (b)   is uniformly convergent on each compact subset;
        (c)   can be repeatedly differentiated term-by-term to give
              Taylor's series expansions for derivatives of $f(z)$ about
              $z = a$.

One procedure for finding Taylor's series expansions for analytic
functions is the familiar (and sometimes tedious) method of taking
successive derivatives.

Example 4.9   The Maclaurin's series for $f(z) = e^z$ is given by $e^z = \sum_{n=0}^{\infty} z^n/n!$, since $f^{(n)}(0) = 1$ for $n = 0, 1, 2, \ldots$. And since e^z is analytic
for all z this expansion is valid for all finite z.

Once we can be sure that a power series representation for $f(z)$ about
$z = a$ (whatever the means of obtaining it) is equivalent to the
Taylor's series for $f(z)$ about $z = a$, we shall be free to take advantage
of any more efficient means for obtaining power series which particular
situations allow.

Example 4.10   By the method of successive derivatives the reader may
verify that in $\{z:|z| < 1\}$ $f(z) = 1/(1 + z)$ has the Maclaurin's expansion
$1/(1 + z) = \sum_{n=0}^{\infty} (-1)^n z^n$, valid for $|z| < 1$. However, by Example 4.5,
we know $1/(1 - w) = \sum_{n=0}^{\infty} w^n$ for $|w| < 1$; by letting $w = -z$ we
obtain $1/(1 + z) = \sum_{n=0}^{\infty} (-1)^n z^n$ for $|z| < 1$.

The next result and its corollaries establish the validity of this approach to obtaining series expansions.

THEOREM 4.7   Hypothesis: The power series $f(z) = \sum_{n=0}^{\infty} a_n(z - a)^n$ has radius of convergence $\rho > 0$.

Conclusion: For $n = 0, 1, 2, \ldots$,

$$a_n = \frac{f^{(n)}(a)}{n!}.$$

(That is, in its disk of convergence the power series is already a Taylor's series about $z = a$ for the analytic function it represents.)

Proof: With $f(z) = \sum_{n=0}^{\infty} a_n(z - a)^n$ in $\{z:|z - a| < \rho\}$, $f(a) = a_0$. Using Theorem 4.3 repeatedly we can see:

$$f'(z) = \sum_{n=1}^{\infty} na_n(z - a)^{n-1}$$

for $|z - a| < \rho$, and $f'(a) = a_1$;

$$f''(z) = \sum_{n=2}^{\infty} n(n - 1)a_n(z - a)^{n-2}$$

for $|z - a| < \rho$, and $f''(a) = 2a_2$. And for any positive integer k,

$$f^{(k)}(z) = \sum_{n=k}^{\infty} n(n - 1)\cdots(n - k + 1)a_n(z - a)^{n-k}$$

for $|z - a| < \rho$, and $f^{(k)}(a) = k!\, a_k$.

Since within their disks of convergence power series are the Taylor's series of the analytic function they represent, if two power series in powers of $(z - a)$ represent the same analytic function in some neighborhood of $z = a$, they must literally be the same power series in the sense that the coefficients of like powers of $(z - a)$ must be the same. Let us state this fact more formally but recognize that it is an obvious consequence of Theorem 4.7.

Corollary 4.1   Hypotheses: The power series $\sum_{n=0}^{\infty} a_n(z - a)^n$ and $\sum_{n=0}^{\infty} b_n(z - a)^n$ have radii of convergence $\rho_1 > 0$ and $\rho_2 > 0$, respectively, and represent the same analytic function in the disk $\{z:|z - a| < \min(\rho_1, \rho_2)\}$.

Conclusion: $a_n = b_n$ for $n = 0, 1, 2, \ldots$.

Another consequence of the Taylor's series approach to analytic functions is worth pointing out.

Corollary 4.2   Hypotheses: Both $f(z)$ and $g(z)$ are analytic in the disk $\{z : |z - a| < r\}$, and for $n = 0, 1, 2, \ldots, f^{(n)}(a) = g^{(n)}(a)$.

Conclusion: $f(z) = g(z)$ for $|z - a| < r$.

Outline of Proof: Note that $F(z) = f(z) - g(z)$ is analytic in the disk $\{z : |z - a| < r\}$; write the Taylor's series for $F(z)$ valid in this disk.

PROBLEMS

4.7   Verify in two ways that for all finite z

$$\exp(z - 1) = \sum_{n=0}^{\infty} \frac{(z - 1)^n}{n!}.$$

4.8   Find the Maclaurin's series for $f(z) = \text{Log}(1 - z)$ in two ways. In what region is this expansion valid?

4.9   (a)   Find the Maclaurin's series for $1/(1 + z^2)$ valid for $|z| < 1$.
   (b)   Use this series to obtain a Maclaurin's series for the principal branch of $\tan^{-1}z$ (that is, the branch for which $\tan^{-1}0 = 0$) valid for $|z| < 1$.

4.10   Suppose $f(z)$ is analytic in $\{z : |z| < r\}$, real-valued when z is real, and for $-r < x < r$, $f(x) = \sum_{n=0}^{\infty} a_n x^n$. Show that $f(z) = \sum_{n=0}^{\infty} a_n z^n$ for $|z| < r$.

Use this fact to write at least the first four terms of the Maclaurin's series for each of the following functions.
   (a) $\sin z$
   (b) $\cos z$
   (c) $\text{Log}(1 + z)$, $|z| < 1$
   (d) $e^z \text{Log}(1 + z)$, $|z| < 1$
   (e) $e^z \sin z$
   (f) e^{e^z}
   (g) $\text{Log}[(1 + z)/(1 - z)]$, $|z| < 1$

4.11   Find the Taylor's series for $\sinh z$ about $z = \pi i$ and give its region of convergence.

4.12 Find the Taylor's series for $1/z^2$ about $z = -1$. (*Suggestion:* $1/z = 1/[(z + 1) - 1] = -1/[1 - (z + 1)]$; rewrite this now in powers of $(z + 1)$.)

4.13 Find the Maclaurin's series for $f(z) = (z + 1)/(z - 1)$.

Section 4.3 Laurent's Series

Suppose $f(z)$ is analytic in the region $\{z{:}0 \leq r < |z - a| < s\}$. This region is not the region of convergence of a Taylor's series, but we can still hope to represent $f(z)$ in this region by a series of (positive and negative) powers of $(z - a)$. This series, known as a Laurent's series, is not a power series, according to our earlier use of these words, but its relation to power series is very close. In particular, if $f(z)$ actually were analytic in the entire region $\{z{:}|z - a| < s\}$, the Laurent's series for $f(z)$ in $\{z{:}0 \leq r < |z - a| < s\}$ would agree with the Taylor's series for $f(z)$ in this region. In this light we can view Taylor's series as a special case of Laurent's series.

There is no meaningful analogue for Laurent's series in the real case. A partial explanation for this is given by saying that $\{z{:}0 \leq r < |z - a| < s\}$ is, like $\{z{:}|z - a| < s\}$, a domain in the complex plane; but on the real line only the second of the sets $\{x{:}0 \leq r < |x - a| < s\}$ and $\{x{:}|x - a| < s\}$ is a domain.

Laurent's Theorem, which leads to a Laurent's series representation of a function, and its proof are fairly technical, although not difficult. Accordingly, we shall not give a complete proof; instead we shall examine the implications and interpretations of this theorem.

THEOREM 4.8 *Laurent's Theorem*
Hypothesis: $f(z)$ is analytic in the region $D = \{z{:}0 \leq r < |z - a| < s\}$.

Conclusions: Throughout $D, f(z)$ can be represented by

$$f(z) = \sum_{n=-\infty}^{\infty} A_n(z - a)^n* \tag{4.1}$$

where $A_n = (1/2\pi i) \int_C [f(z)/(z - a)^{n+1}]\, dz$, $n = 0, \pm 1, \pm 2, \ldots$ and C is any closed path in D enclosing $z = a$.

---

*By the symbol $\sum_{n=-\infty}^{\infty} a_n$ we mean the formal sum, $\sum_{n=0}^{\infty} a_n + \sum_{n=1}^{\infty} a_{-n}$, of two infinite series. We say $\sum_{n=-\infty}^{\infty} a_n$ is convergent if and only if both the series $\sum_{n=0}^{\infty} a_n$ and $\sum_{n=1}^{\infty} a_{-n}$ are convergent.

At each point of D the series (4.1) is absolutely convergent to $f(z)$, and on each compact subset of D the series (4.1) is uniformly convergent to $f(z)$.

Outline of Proof: The details of the proof rest on the fact that $\sum_{n=0}^{\infty} w^n$ is absolutely convergent for $|w| < 1$ and is uniformly convergent on compact subsets of $\{w:|w| < 1\}$. For any positive integer N, the Nth partial sum of this series is $\sum_{n=0}^{N} w^n = (1 - w^{N+1})/(1 - w)$, $|w| < 1$.

We choose any point z_0 in D and positive number ϵ so that $r + \epsilon < |z_0 - a| < s - \epsilon$. Then $C_1 = \{z:|z - a| = r + \epsilon\}$ and $C_2 = \{z:|z - a| = s - \epsilon\}$ both lie in D and z_0 lies between them. (See Fig. 4.4.) Let L be the intersection of $\{z:r + \epsilon \leq |z - a| \leq s - \epsilon\}$ with any ray from $z = a$ not passing through $z = z_0$. Let P and Q be the endpoints of L, where P is on C_1 and Q is on C_2. The closed path (see the arrows in Fig. 4.4) $K = L \oplus C_2 \oplus (-L) \oplus (-C_1)$ forms the boundary of a simply connected region containing z_0, and $f(z)$ is analytic on and inside K. Using a slightly generalized version of Theorem 3.4 we may write

$$f(z_0) = (1/2\pi i) \int_K \left[\frac{f(z)}{(z - z_0)} \right] dz$$

$$= (1/2\pi i) \int_{C_2} \left[\frac{f(z)}{(z - z_0)} \right] dz$$

$$- (1/2\pi i) \int_{C_1} \left[\frac{f(z)}{(z - z_0)} \right] dz \qquad (4.2)$$

We can deal with each of these integrals separately but similarly.

For $z \in C_2$, $|z - a| > |z_0 - a|$, and $z - z_0 = (z - a)[1 - (z_0 - a)/(z - a)]$. By letting $w = (z_0 - a)/(z - a)$, we have $|w| < 1$ for $z \in C_2$, and an argument like that in the proof of Theorem 4.6 shows that

$$\frac{1}{2\pi i} \int_{C_2} \frac{f(z)}{z - z_0} dz = \sum_{n=0}^{\infty} \left[\frac{1}{2\pi i} \int_{C_2} \frac{f(z)}{(z - a)^{n+1}} dz \right] (z_0 - a)^n \qquad (4.3)$$

When $z \in C_1$, $|z_0 - a| > |z - a|$, and $z - z_0 = -(z_0 - a)[1 - (z - a)/(z_0 - a)]$. Here we let $w = (z - a)/(z_0 - a)$, so that $|w| < 1$ for $z \in C_1$. The same kind of argument leads to the equation

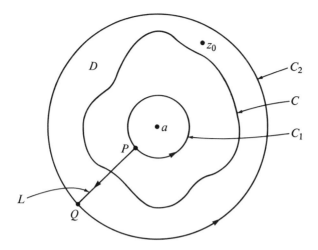

Figure 4.4

$$\frac{1}{2\pi i}\int_{C_1}\frac{f(z)}{z-z_0}\,dz = \sum_{k=1}^{\infty}\left[\frac{1}{2\pi i}\int_{C_1}f(z)(z-a)^{k-1}\,dz\right](z_0-a)^{-k}$$

$$= \sum_{n=-1}^{-\infty}\left[\frac{1}{2\pi i}\int_{C_1}\frac{f(z)}{(z-a)^{n+1}}\,dz\right](z_0-a)^n \qquad (4.4)$$

If C is any closed path in D enclosing $z = a$, we can write

$$\frac{1}{2\pi i}\int_{C_2}\frac{f(z)}{(z-a)^{n+1}}\,dz = \frac{1}{2\pi i}\int_{C}\frac{f(z)}{(z-a)^{n+1}}\,dz, \qquad (4.5)$$

$$n = 0, 1, 2, \dots$$

and

$$\frac{1}{2\pi i}\int_{C_1}\frac{f(z)}{(z-a)^{n+1}}\,dz = \frac{1}{2\pi i}\int_{C}\frac{f(z)}{(z-a)^{n+1}}\,dz, \qquad (4.6)$$

$$n = -1, -2, -3, \dots.$$

(Equation (4.6) is a consequence of Theorem 3.3, since for $n = -1, -2, -3, \dots$, $f(z)/(z-a)^{n+1}$ is analytic in D. Equation (4.5) would require generalizations of the theorems in Chapter 3 for its proof.)

By using equations (4.3)–(4.6) in equation (4.2), we have

$$f(z_0) = \sum_{n=0}^{\infty} \left[\frac{1}{2\pi i} \int_C \frac{f(z)}{(z-a)^{n+1}} \, dz \right] (z_0 - a)^n$$

$$+ \sum_{n=-1}^{-\infty} \left[\frac{1}{2\pi i} \int_C \frac{f(z)}{(z-a)^{n+1}} \, dz \right] (z_0 - a)^n$$

$$= \sum_{n=-\infty}^{\infty} \left[\frac{1}{2\pi i} \int_C \frac{f(z)}{(z-a)^{n+1}} \, dz \right] (z_0 - a)^n,$$

which is equation (4.1).

The absolute convergence of the series (4.1) at points of D and its uniform convergence on compact subsets of D are established by arguments based on the steps leading to equations (4.3) and (4.4).

The Laurent's series representation (4.1) for $f(z)$ in the region $\{z : 0 \leq r < |z - a| < s\}$ is somewhat easier to find and use than it may appear at first. Before we examine some examples to support this, we make some comments on what we have.

  (1) In general it is not practical to calculate the coefficients A_n directly by integration. In fact, after finding a Laurent's expansion by other means, we use the coefficients thus obtained to evaluate integrals of the form $(1/2\pi i) \int_C [f(z)/(z-a)^{n+1}] \, dz$.

  (2) If $f(z)$ is actually analytic in $\{z : |z - a| < s\}$, then $f(z)/(z-a)^{n+1}$ is analytic there for $n = -1, -2, -3, \ldots$ and $A_n = 0$ for $n = -1, -2, -3, \ldots$. That is, the Laurent's series reduces to the Taylor's series.

  (3) Our primary use of Laurent's series will be for functions $f(z)$ analytic in $\{z : 0 < |z - a| < s\}$ (analytic near $z = a$ but perhaps not analytic at $z = a$ itself). The number of negative powers of $(z - a)$ present in the Laurent's expansion for $f(z)$ in $\{z : 0 < |z - a| < s\}$ will serve as a measure of how "non-analytic" $f(z)$ is at $z = a$.

It is important to know that a function has only one Laurent's series expansion, so we state the following result for completeness. It contains Theorem 4.7 as a special case.

THEOREM 4.9   Hypothesis: The series $\sum_{n=-\infty}^{\infty} A_n(z-a)^n$ represents a function $f(z)$, analytic in the region $\{z : 0 \leq r < |z - a| < s.\}$

Conclusion: For $n = 0, \pm1, \pm2, \ldots$

$$A_n = \frac{1}{2\pi i} \int_C \frac{f(z)}{(z-a)^{n+1}} \, dz,$$

where C is any closed path in $\{z : 0 \leq r < |z - a| < s\}$ enclosing $z = a$. (That is, the series is the Laurent's series for $f(z)$ in the region $\{z : 0 \leq r < |z - a| < s\}$.)

Example 4.11   Laurent's Theorem does not apply to $f(z) = \text{Log } z$ in any annular region about $z = 0$, because no branch of the logarithm function, $\log z$, is analytic at each point of an annular region about $z = 0$.

Example 4.12   The function $f(z) = 1/(z-1)^2$ is analytic for all z, $|z - 1| > 0$. The expression for $f(z)$ is already the Laurent's series for $f(z)$ valid in $\{z : 0 < |z - 1| < r\}$ for any number $r > 0$.

Example 4.13   Let $f(z) = 1/(z^2 - 3z + 2)$; $f(z)$ is analytic for all z except $z = 1$, $z = 2$. We shall find three essentially different expansions of $f(z)$ in powers of z. In terms of regions centered at $z = 0$, Laurent's Theorem implies $f(z)$ has a Laurent's series in (a) $|z| < 1$; (b) $1 < |z| < 2$; (c) $2 < |z| < r$ for any number $r > 2$.

Write $$f(z) = \frac{1}{(z-2)(z-1)} = \frac{1}{z-2} - \frac{1}{z-1}.$$

(a) $|z| < 1$ Here the expansion in powers of z will be the Maclaurin's series for $f(z)$,

$$\frac{1}{z-2} = -\frac{1}{2}\left(\frac{1}{1 - \frac{z}{2}}\right) = -\frac{1}{2}\sum_{n=0}^{\infty} \frac{z^n}{2^n},$$

valid for $|z/2| < 1$, or $|z| < 2$.

$$\frac{1}{z-1} = -\left(\frac{1}{1-z}\right) = -\sum_{n=0}^{\infty} z^n,$$

valid for $|z| < 1$.

For $|z| < 1$

$$f(z) = -\frac{1}{2}\sum_{n=0}^{\infty}\frac{z^n}{2^n} + \sum_{n=0}^{\infty}z^n = \sum_{n=0}^{\infty}\left(1 - \frac{1}{2^{n+1}}\right)z^n.$$

(b) $\underline{1 < |z| < 2}$

$1/(z-2) = -\frac{1}{2}\sum_{n=0}^{\infty}z^n/2^n$, valid for $|z| < 2$, as in (a).

$$\frac{1}{z-1} = \frac{1}{z}\left(\frac{1}{1-\frac{1}{z}}\right) = \frac{1}{z}\sum_{n=0}^{\infty}\frac{1}{z^n} = \sum_{n=0}^{\infty}\frac{1}{z^{n+1}},$$

valid for $|1/z| < 1$, or $|z| > 1$. Then for $1 < |z| < 2$,

$$f(z) = -\frac{1}{2}\sum_{n=0}^{\infty}\frac{z^n}{2^n} - \sum_{-\infty}^{-1}z^n = \sum_{n=-\infty}^{\infty}A_nz^n,$$

where $A_n = \begin{cases} -\dfrac{1}{2^{n+1}} \text{ for } n \geq 0. \\ -1 \text{ for } n < 0 \end{cases}$

(c) $\underline{2 < |z| < r \text{ for any } r > 2}$

$$\frac{1}{z-2} = \frac{1}{z}\left(\frac{1}{1-\frac{2}{z}}\right) = \frac{1}{z}\sum_{n=0}^{\infty}\left(\frac{2}{z}\right)^n = \sum_{n=0}^{\infty}\frac{2^n}{z^{n+1}},$$

valid for $|2/z| < 1$, or $|z| > 2$. $1/(z-1) = \sum_{n=0}^{\infty}1/z^{n+1}$, valid for $|z| > 1$, as in (b). For $2 < |z|$,

$$f(z) = \sum_{n=0}^{\infty}\frac{2^n}{z^{n+1}} - \sum_{n=0}^{\infty}\frac{1}{z^{n+1}} = \sum_{n=0}^{\infty}\frac{2^n - 1}{z^{n+1}}$$

$$= \sum_{n=-\infty}^{-1}(2^{-n-1} - 1)z^n = \sum_{n=-\infty}^{\infty}A_nz^n,$$

where $A_n = \begin{cases} 0 & n \geq 0 \\ (2^{-n-1} - 1) & n < 0 \end{cases}$

Example 4.14   Let $f(z) = e^z/z^3$. Since $e^z = \sum_{n=0}^{\infty}z^n/n!$ for all finite z, while $1/z^3$ is analytic for $|z| > 0$, the Laurent's series for $f(z)$ valid in the region $\{z: 0 < |z| < r\}$ for any number $r > 0$ will be

$$f(z) = \frac{e^z}{z^3} = \sum_{n=0}^{\infty}\frac{z^{n-3}}{n!} = \sum_{n=-3}^{\infty}\frac{z^n}{(n+3)!}$$

PROBLEMS

4.14 Write Laurent's series for $f(z) = 1/[z(z^2 + 4)]$
  (a) valid for $0 < |z| < 2$;
  (b) valid for $2 < |z|$;
  (c) valid for $0 < |z - 2i| < 2$;
  (d) valid for $4 < |z - 2i|$.

4.15 Find Laurent's series to represent the function $1/[(z^2 + 1)(z + 2)]$ in the regions
  (a) $|z| < 1$;
  (b) $1 < |z| < 2$;
  (c) $2 < |z|$;
  (d) $0 < |z + 2| < 5^{1/2}$.

4.16 Find Laurent's series valid for $0 < |z|$ for the functions
  (a) $f(z) = \exp(1/z)$;
  (b) $f(z) = \exp(-1/z^2)$.

4.17 If $f(z) = \sin(1/z)$ find a Laurent's series valid in a region $\{z : 0 < |z| < r\}$ for some positive number r. What is the largest value r may have?

4.18 If $f(z) = (z - 1)^2 + (z - 1)^{-2}$, what is the Laurent's expansion for $f(z)$ valid for $|z - 1| > 0$?

4.19 Use Laurent's Theorem to verify that for $|z| > 0$

$$\cosh(z + z^{-1}) = A_0 + \sum_{n=1}^{\infty} A_n(z^n + z^{-n}),$$

where $\qquad A_n = (1/2\pi) \int_0^{2\pi} \cos(n\theta) \cosh(2 \cos \theta)d\theta,$

$n = 0, 1, 2, \ldots.$

4.20 If c is a real constant, verify that for $|z| > 0$

$$\exp\left[\frac{c}{2}\left(z - \frac{1}{z}\right)\right] = \sum_{n=-\infty}^{\infty} A_n z^n,$$

$$\text{where } \begin{cases} A_n = (1/2\pi)\int_0^{2\pi} \cos[n\theta - c\sin\theta]d\theta \\ A_{-n} = (-1)^n A_n \end{cases}, \; n = 0, 1, 2, \ldots.$$

(For each integer n, A_n is the value at $x = c$ of the Bessel's function, $J_n(x)$, of the first kind of order n.)

Section 4.4   Classification of Isolated Singularities

If $f(z)$ is a function which is analytic at some points and not analytic at others, it is common to call the points at which $f(z)$ is not analytic *singularities*, or *singular points*, of $f(z)$. If $z = a$ is a singularity of $f(z)$, but for some number $r > 0$, $f(z)$ is analytic in the region $\{z : 0 < |z - a| < r\}$, we call $z = a$ an *isolated singularity of* $f(z)$. With the aid of Laurent's series we can classify the isolated singular points of otherwise analytic functions into three kinds. By examining the Laurent's series information in each case we shall be able to find other distinguishing features of each kind of isolated singularity. This information will help us later to identify isolated singularities without resorting to Laurent's series.

Suppose $f(z)$ is analytic for $0 < |z - a| < r$ but we have no information about $f(z)$ at $z = a$. If $f(z) = \sum_{n=-\infty}^{\infty} A_n(z - a)^n$ for $0 < |z - a| < r$, we call $\sum_{n=-\infty}^{-1} A_n(z - a)^n$ the *principal part of* $f(z)$ *at* $z = a$, and we write

$$P(f,a) = \sum_{n=-\infty}^{-1} A_n(z - a)^n,$$

so that

$$f(z) = P(f,a) + \sum_{n=0}^{\infty} A_n(z - a)^n.$$

Note that $f(z) - P(f,a) = \sum_{n=0}^{\infty} A_n(z - a)^n$ is in the form of a Taylor's series; we could show from Laurent's Theorem that $f(z) - P(f,a)$ is analytic for $|z - a| < r$. So $P(f,a)$ will tell us about the "non-analytic" behavior of $f(z)$ at $z = a$.

Case I $P(f,a) = 0$.

Here the Laurent's expansion for $f(z)$ in the region $\{z : 0 < |z - a| < r\}$ is actually a Taylor's expansion of $f(z) = \sum_{n=0}^{\infty} A_n(z - a)^n$. If we supply the value A_0 for $f(a)$, then $f(z)$ will be analytic for $|z - a| < r$, including the point $z = a$.

For this reason, when $P(f,a) = 0$ we call $z = a$ a *removable singularity* for $f(z)$, and we consider $f(z)$ to be analytic at $z = a$ with $f(a)$ defined to be A_0.

Example 4.15 $f(z) = \sin z/z$ is analytic for $|z| > 0$. But

$$f(z) = \frac{1}{z} \sum_{n=0}^{\infty} \frac{(-1)^n z^{2n+1}}{(2n + 1)!} = \sum_{n=0}^{\infty} \frac{(-1)^n z^{2n}}{(2n + 1)!} = 1 - \frac{z^2}{3!} + \frac{z^4}{5!} - \cdots$$

for $|z| > 0$, and $P[(\sin z/z),0] = 0$. Then $z = 0$ is a removable singularity for $\sin z/z$, where we define $f(0) = [(\sin z)/(z)]_{z=0} = 1$.

Case II $P(f,a)$ has finitely many terms.
   Then for some positive integer m, $f(z) = \sum_{n=-m}^{\infty} A_n(z - a)^n$, $P(f,a) = \sum_{n=-m}^{-1} A_n(z - a)^n$, where $A_{-m} \neq 0$, and $A_n = 0$ for $n < -m$.
   In this case we call $z = a$ a *pole of order m* for $f(z)$. (A reason for this choice of words will emerge later.) A pole of order one is frequently called a *simple pole*.

Example 4.16   In Example 4.14 we found that

$$\frac{e^z}{z^3} = \frac{1}{z^3} + \frac{1}{z^2} + \frac{1}{2z} + \sum_{n=0}^{\infty} \frac{z^n}{(n + 3)!} \quad \text{for} \quad |z| > 0.$$

Thus $P(e^z/z^3,0) = 1/z^3 + 1/z^2 + 1/2z$, and e^z/z^3 has a pole of order 3 at $z = 0$.

Case III $P(f,a)$ has infinitely many terms.
   In this case we call $z = a$ an *essential singular point* (ESP) for $f(z)$.

Example 4.17   From problem 4.16(a) we have

$$\exp\left(\frac{1}{z}\right) = \sum_{n=-\infty}^{0} \frac{z^n}{(-n)!} \quad \text{for} \quad |z| > 0.$$

In this case $\exp(1/z) = P[\exp(1/z),0] + 1$, and $\exp(1/z)$ has an ESP at $z = 0$.

Each of the three cases above can be described also in terms of the behavior of $f(z)$ as z approaches a.

In Case I we see that when $P(f,a) = 0$, $\lim_{z \to a} f(z)$ will have a finite value A_0. Conversely, if $\lim_{z \to a} f(z)$ exists, it will be true that $P(f,a) = 0$.

In Case II let us suppose $f(z)$ has a pole of order m at $z = a$. Then for $0 < |z - a| < r$

$$f(z) = \sum_{n=-m}^{\infty} A_n(z - a)^n$$

$$= (z - a)^{-m} \sum_{n=-m}^{\infty} A_n(z - a)^{n+m}$$

$$= (z - a)^{-m} \sum_{k=0}^{\infty} A_{k-m}(z - a)^k$$

$$= (z - a)^{-m}\phi(z),$$

where $\phi(z) = \sum_{k=0}^{\infty} A_{k-m}(z - a)^k$ is analytic for $|z - a| < r$, and $\phi(a) = A_{-m} \neq 0$. (Why?)

We can state this as a theorem.

THEOREM 4.10 Hypothesis: $f(z)$ is analytic in $\{z : 0 < |z - a| < r\}$.

Conclusion: $f(z)$ has a pole of order m at $z = a$ if and only if $f(z) = \phi(z)(z - a)^{-m}$, where $\phi(z)$ is analytic at $z = a$ and $\phi(a) \neq 0$.

Example 4.18 Let $f(z) = \cos z/[z^2(z^2 + 1)]$; $f(z)$ has singularities at $z = 0, i, -i$.

If we write $f(z) = \cos z/[z^2(z - i)(z + i)]$, we see that $f(z)$ has a pole of order 2 at $z = 0$ because $f(z) = [\cos z/(z - i)(z + i)]/z^2$, where $\cos z/(z - i)(z + i)$ is analytic at $z = 0$ and different from zero there. Also $f(z)$ has a pole of order 1 (a simple pole) at $z = i$, because $f(z) = [\cos z/z^2(z + i)]/(z - i)$, where $\cos z/[z^2(z + i)]$ is analytic and nonzero at $z = i$. Similarly, $f(z)$ has a simple pole at $z = -i$.

Theorem 4.10 has the following consequence.

Corollary 4.3 Hypothesis: $f(z)$ has a pole of order m at $z = a$.

Conclusion: $\lim_{z \to a} |f(z)| = +\infty$.

The choice of the word "pole" in Case II seems reasonable if one tries to visualize the surface $w = |f(x + iy)|$ over the complex plane near the point $z = x + iy = a$.

For another perspective on poles we introduce another term.

Definition   Suppose $f(z)$ is analytic at $z = a$ and $f(z) = \sum_{n=0}^{\infty} a_n(z - a)^n$ for $|z - a| < r$. We say $f(z)$ has a *zero of order m at $z = a$* if $a_0 = a_1 = \cdots = a_{m-1} = 0$ but $a_m \neq 0$. We frequently call a zero of order 1 a *simple zero.*

If $f(z)$ has a zero of order m at $z = a$, then

$$f(z) = (z - a)^m \sum_{n=m}^{\infty} a_n(z - a)^{n-m}$$

$$= (z - a)^m \sum_{k=0}^{\infty} a_{m+k}(z - a)^k$$

$$= (z - a)^m \, \phi(z),$$

where $\phi(z)$ is analytic at $z = a$ and $\phi(a) = a_m \neq 0$. Conversely, if $f(z) = (z - a)^m \phi(z)$, where $\phi(z)$ is analytic and nonzero at $z = a$, then $f(z)$ has a zero of order m at $z = a$.

Example 4.19   If $f(z) = \sin(z^2)$, $f(z) = \sum_{n=0}^{\infty} [(-1)^n(z^2)^{2n+1}]/(2n + 1)!$ and $f(z)$ has a zero of order 2 at $z = 0$.

There is an obvious similarity between these remarks and Theorem 4.10. We can summarize it in this result.

THEOREM 4.11   Hypothesis: $f(z)$ has a pole of order m at $z = a$.

Conclusion: $1/f(z)$ is analytic and has a zero of order m at $z = a$.

Proof: By Theorem 4.10, $f(z) = (z - a)^{-m} \phi(z)$, where $\phi(z)$ is analytic at $z = a$, and $\phi(a) \neq 0$.

Then for some number $\delta > 0$, $\phi(z) \neq 0$ when $|z - a| < \delta$, $1/f(z) = (z - a)^m/\phi(z)$, where $1/\phi(z)$ is analytic and nonzero at $z = a$. That is, $1/f(z)$ has a zero of order m at $z = a$.

As we saw in Case III, if $f(z)$ has an ESP at $z = a$, the behavior of $f(z)$ near $z = a$ is extremely complicated. We can indicate this by stating what is known as the *Casorati-Weierstrass Theorem.* (We will outline the proof of this theorem in problem 4.31.)

THEOREM 4.12   Hypotheses: $f(z)$ has an ESP at $z = a$, and c is *any* complex number.

Conclusion: There exists a sequence $\{z_n\}$, with $\lim_{n \to \infty} z_n = a$, such that $\lim_{n \to \infty} f(z_n) = c$.

Suppose $f(z)$ has an isolated singularity at $z = a$. If we can find a sequence of points $\{z_n\}$ with $\lim_{n\to\infty} z_n = a$ and $\lim_{n\to\infty} f(z_n) = b$, and another sequence of points $\{w_n\}$ with $\lim_{n\to\infty} w_n = a$ and $\lim_{n\to\infty} f(w_n) = c$, where $b \neq c$, then:

   (1) $f(z)$ does not have a removable singularity at $z = a$. (Why?)
   (2) $f(z)$ does not have a pole at $z = a$. (Why?)
   (3) $f(z)$ must have an ESP at $z = a$.

Example 4.20 Let $f(z) = \sin(1/z)$.
  If $z_k = 1/\pi k$, $k = 1, 2, 3, \ldots$, $\lim_{k\to\infty} z_k = 0$,

$$\lim_{k\to\infty} \sin\left(\frac{1}{z_k}\right) = \lim_{k\to\infty} \sin(\pi k) = 0;$$

if $w_k = 2/[\pi(2\kappa + 1)]$, $k = 2, 4, 6, \ldots$,

$$\lim_{k\to\infty} w_k = 0, \lim_{k\to\infty} \sin\left(\frac{1}{w_k}\right) = \lim_{k\to\infty} \sin[\pi(k + \tfrac{1}{2})] = 1.$$

Thus $\sin(1/z)$ has an ESP at $z = 0$.

PROBLEMS

4.21 For each function below list all the points where the function has a pole or a zero and give its order.
   (a) $z/(z + 1)$
   (b) $e^z/(z^2 + 1)$
   (c) $e^z - 1$
   (d) $1/(e^z - 1)$
   (e) $(z^3 - 1)/(e^z - 2)$
   (f) $(z - z^{-1})^2$

4.22 We know $\sin z = 0$ if and only if $z = k\pi$, k an integer.
   (a) Show that each $z = k\pi$ is a zero of order 1 for $\sin z$.
   (b) Show that each $z = k\pi$ is a simple pole for $\csc z$.

4.23 Is every zero for $\cos z$ a zero of order 1?

4.24 At what points does $\tan z$ have a pole, and what is the order of each of its poles?

4.25 List the points where $1/(z \sin z)$ has a pole and give the order of the pole at each such point.

4.26 Suppose $f(z)$ has a pole of order m at $z = a$, $g(z)$ has a pole of order n at $z = a$. Verify that $f(z)/g(z)$

   (a)   has a zero of order $n - m$ at $z = a$ if $m < n$;

   (b)   has a pole of order $m - n$ at $z = a$ if $m > n$;

   (c)   is analytic and nonzero at $z = a$ if $m = n$.

4.27 Suppose $f(z)$ is analytic at $z = a$ and $f(a) = 0$.

   (a)   Show that for some number $\delta > 0$ either $f(z) \equiv 0$ for $|z - a| < \delta$ or $f(z) \neq 0$ for $0 < |z - a| < \delta$.

   (b)   If $f(z)$ is not identically zero for $|z - a| < \delta$, then $1/f(z)$ has a pole at $z = a$ whose order is the order of the zero of $f(z)$ at $z = a$. (A converse to Theorem 4.11.)

4.28 (a)   Suppose $f(z)$ and $g(z)$ are analytic at $z = a$ while $f(a) = g(a) = 0$. Prove that if $\lim_{z \to a} f'(z)/g'(z)$ exists, so does $\lim_{z \to a} f(z)/g(z)$ and the limits are equal. (L' Hospital's Rule.)

   (b)   Evaluate $\lim_{z \to \pi} [(z - \pi)^2 \cos z]/\sin^2 z$.

4.29 Suppose $f(z)$ is analytic for $0 < |z - a| < r$ and there is a sequence of points $\{z_n\}$ with $\lim_{n \to \infty} z_n = a$ such that $f(z_n) = c$, a finite constant, for $n = 1, 2, 3, \ldots$. Prove that $f(z)$ has an ESP at $z = a$. (*Hint:* Look at $g(z) = f(z) - a$, which is analytic for $0 < |z - a| < r$; apply problem 4.27(b) to $g(z)$ to show $g(z)$ is not analytic at $z = a$. Then show $g(z)$ does not have a pole at $z = a$.)

4.30 Suppose $f(z)$ is analytic, and $|f(z)| \leq M$, for $0 < |z - a| < r$. Prove that $z = a$ is a removable singularity for $f(z)$.

4.31 Suppose $f(z)$ has an ESP at $z = a$. An indirect proof for the Casorati-Weierstrass Theorem can be given by justifying each of the following statements.

   (a)   If the Casorati-Weierstrass Theorem is not true, there must be a complex number c_0, and positive numbers ϵ and δ, such that $|f(z) - c_0| \geq \epsilon$ whenever $0 < |z - a| < \delta$.

   (b)   Then $g(z) = 1/[f(z) - c_0]$, which is analytic for $0 < |z - a| < \delta$, must have a removable singularity at $z = a$, and we can define $b = g(a)$ so as to make $g(z)$ analytic at $z = a$.

   (c)   If $b \neq 0$, then $f(z)$ has a removable singularity at $z = a$; if $b = 0$, then $f(z)$ has a pole at $z = a$.

   Either conclusion in (c) is a contradiction, so the conclusion of the Casorati-Weierstrass Theorem must hold for $f(z)$.

5

Residues and
Evaluation of Integrals

By exploiting a chain of results derived from the Cauchy Integral Theorem and the classification of isolated singularities derived from Laurent's Theorem, we can generate an exceedingly rich and interesting collection of techniques for evaluating line integrals and improper definite integrals. This chapter can present only a sample of these techniques. The basis for them is roughly this: If C is a closed path on and within which $f(z)$ is analytic except for isolated singularities, then $\int_C f(z)\,dz$ is a number which depends only on the number and nature of those singularities of $f(z)$ enclosed in C. In effect, we shall evaluate integrals about closed paths by examining isolated singularities.

Section 5.1   Residues

Suppose $f(z)$ is analytic in the region $\{z : 0 < |z - a| < r\}$ and has the Laurent's expansion $\sum_{n=-\infty}^{\infty} a_n(z - a)^n$ there. For any closed path C in this region enclosing $z = a$, the convergence properties of the Laurent's series permit us to write

$$\int_C f(z)\, dz = \sum_{n=-\infty}^{\infty} a_n \int_C (z - a)^n\, dz$$

$$= 2\pi i\, a_{-1}. \quad \text{(See problem 3.4.)}$$

That is, when we integrate $f(z)$ about any closed path enclosing the isolated singularity $z = a$ (and no other singularities of $f(z)$), all that remains is a multiple of one coefficient of the Laurent's expansion of $f(z)$, the coefficient of $(z - a)^{-1}$. This implies that to evaluate $(1/2\pi i) \int_C f(z)\, dz$ we derive the Laurent's series for $f(z)$ about $z = a$ and extract only one coefficient. An obvious improvement on this extravagant method would be to acquire the coefficient we want without obtaining the whole Laurent's series.

Definition   Let $f(z)$ be analytic in $\{z : 0 < |z - a| < r\}$. We define the *residue of $f(z)$ at $z = a$* to be the coefficient a_{-1} of $(z - a)^{-1}$ in the Laurent's series for $f(z)$ about $z = a$. And we write

$$\text{Res}[f, a] = a_{-1}$$

If $f(z)$ is analytic at $z = a$, or if $z = a$ is a removable singularity for $f(z)$, we see that $\text{Res}[f, a] = 0$.

Example 5.1   Let $f(z) = e^z/(z - 2)^2$ and C be the closed path defined by $z = 2 + \exp(i\theta)$, $0 \le \theta \le 2\pi$. The function $f(z)$ has a pole of order 2 at $z = 2$, and

$$f(z) = \frac{e^z}{(z - 2)^2} = (z - 2)^{-2}\left[e^2 \sum_{n=0}^{\infty} \frac{(z - 2)^n}{n!}\right]$$

$$= \frac{e^2}{(z - 2)^2} + \frac{e^2}{z - 2} + e^2 \sum_{n=2}^{\infty} \frac{(z - 2)^{n-2}}{n!};$$

so $\text{Res}[e^z/(z - 2)^2, 2] = e^2$, and

$$(1/2\pi i)\int_C \left[\frac{e^z}{(z-2)^2}\right] dz = e^2.$$

Note that we can also evaluate this integral as the derivative of e^z at $z = 2$ by the Cauchy integral formula with $n = 1$.

If $f(z)$ is analytic on and inside a closed path C, except for a pole of order m at a point $z = a$ inside C, then we can represent $f(z)$ by $(z - a)^{-m}\phi(z)$, where $\phi(z)$ is analytic at $z = a$ and $\phi(a) \neq 0$. We can evaluate $(1/2\pi i)\int_C f(z)\, dz$ in two ways:

$$(1/2\pi i)\int_C f(z)\, dz = \begin{cases} \text{Res}[f,a], \text{ or} \\ \dfrac{\phi^{(m-1)}(a)}{(m-1)!} \end{cases}$$

the second evaluation being a consequence of the Cauchy integral formulas. The first evaluation, involving the use of a Laurent's series for $f(z)$ about $z = a$, may in some cases take more time to obtain.

In problems 3.14 and 3.15 we have seen that if $f(z)$ has more than one pole inside C, some algebraic manipulations are required in order to apply the Cauchy integral formulas in evaluating $(1/2\pi i)\int_C f(z)\, dz$. And if $f(z)$ has an ESP inside C it will not be possible to evaluate $(1/2\pi i)\int_C f(z)\, dz$ at all through the Cauchy integral formulas. If, however, methods for calculating residues are available which do not require deriving many Laurent's series, the following theorem will persuade us that the residue method for evaluating integrals over closed paths is very useful and efficient.

THEOREM 5.1   *Cauchy Residue Theorem* Hypotheses: C is a closed path and $f(z)$ is analytic on and inside C except at the points a_1, a_2, $\ldots$, a_n inside C.

Conclusion:

$$(1/2\pi i)\int_C f(z)\, dz = \sum_{k=1}^n \text{Res}[f,a_k].$$

Outline of Proof: We can choose a number $\delta > 0$ so small that the circles $C_k\{z:|z - a_k| = \delta\}$, $k = 1, 2, \ldots, n$, about the points $a_1, a_2, \ldots, a_n$

do not intersect each other or C. For each k, $1 \leq k \leq n$, let L_k be a path from C to C_k, chosen so that no two paths intersect each other and L_k intersects no circle but C_k. If P is any point on C not on any L_k, we let T be a closed curve described as follows: starting at P we go along C

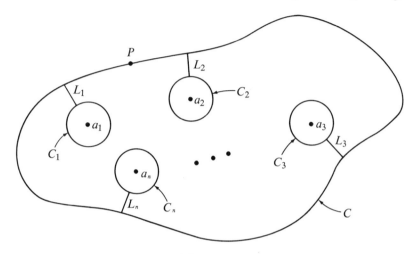

Figure 5.1

in the positive direction until we come to a point on some L_k; we go inside C along L_k to a point on C_k; we go about C_k in the negative direction back to L_k and back out along L_k to C; we then go along C in the positive direction until we come to a point on another L_j where we repeat the procedure above; after n such "excursions" inside C we return to P along C in the positive direction.

Since $f(z)$ is analytic on and inside T, $(1/2\pi i) \int_T f(z) \, dz = 0$. Also, from the properties of these integrals and the curve T,

$$(1/2\pi i) \int_T f(z) \, dz = (1/2\pi i) \int_C f(z) \, dz + \sum_{k=1}^{n} (1/2\pi i) \int_{-C_k} f(z) \, dz.$$

(What has happened to the integrals along $L_1, L_2, \ldots, L_k$?) Thus

$$(1/2\pi i) \int_C f(z) \, dz = \sum_{k=1}^{n} (1/2\pi i) \int_{C_k} f(z) \, dz.$$

Our earlier remarks have shown that

$$(1/2\pi i) \int_{C_k} f(z) \, dz = \text{Res}[f,a_k], \quad k = 1, 2, \ldots, n.$$

It follows that

$$(1/2\pi i) \int_C f(z) \, dz = \sum_{k=1}^{n} \text{Res}[f,a_k].$$

Example 5.2 Let C be the circle $\{z:|z| = 2\}$ and $f(z) = (\exp z)/[(z - 1)^3 \sin z]$. Except for a simple pole at $z = 0$ and a pole of order 3 at $z = 1, f(z)$ is analytic on and inside C, so

$$(1/2\pi i) \int_C f(z) \, dz = \text{Res}[f,0] + \text{Res}[f,1].$$

We will later ask the reader to find these residues (problem 5.1).

Instead of finding these residues directly, let us turn to the problem of using Laurent's series to derive rules for calculating residues.

Suppose $f(z)$ has a pole of order m at $z = a$. Then for some number $r > 0, f(z) = (z - a)^{-m}\phi(z)$ for $0 < |z - a| < r$, where $\phi(z)$ is analytic at $z = a$ and $\phi(a) \neq 0$. We can write $\phi(z)$ in a Taylor's series at $z = a$,

$$\phi(z) = \sum_{n=0}^{\infty} \frac{\phi^{(n)}(a)}{n!}(z - a)^n,$$

and so obtain the Laurent's series for $f(z)$ in $\{z:0 < |z - a| < r\}$,

$$f(z) = \sum_{n=0}^{\infty} \frac{\phi^{(n)}(a)}{n!}(z - a)^{n-m},$$

from which we see that

$$\text{Res}[f,a] = \frac{\phi^{(m-1)}(a)}{(m - 1)!},$$

where $\phi(z) = (z - a)^m f(z)$.

If a is a simple pole for $f(z)$ (that is, $m = 1$) we have $\text{Res}[f,a] = \phi(a)$, where $\phi(z) = (z - a)f(z)$.

One situation arises frequently enough to make it worthwhile to formulate an additional rule for calculating residues.

THEOREM 5.2 Hypotheses: $f(z) = g(z)/h(z)$, where $g(z)$ and $h(z)$ are analytic at $z = a$, $g(a) \neq 0$, and $h(z)$ has a simple zero at $z = a$.

Conclusions: $f(z)$ has a simple pole at $z = a$, and $\text{Res}[f,a] = g(a)/h'(a)$.

Proof: We can write $h(z) = (z - a)k(z)$, where $k(z)$ is analytic at $z = a$ and $k(a) \neq 0$. Then $f(z) = g(z)/[(z - a)k(z)] = (z - a)^{-1}[g(z)/k(z)]$, where $g(z)/k(z)$ is analytic at $z = a$, and $g(a)/k(a) \neq 0$. Thus $f(z)$ has a simple pole at $z = a$.

By the rule derived above, with $\phi(z) = g(z)/k(z)$, $\text{Res}[f,a] = g(a)/k(a)$. Now $h(z) = (z - a)k(z)$, and $k(a) = \lim_{z \to a}[h(z)/(z - a)] = \lim_{z \to a}\{[h(z) - h(a)]/(z - a)\} = h'(a)$. Thus $\text{Res}[f,a] = g(a)/h'(a)$.

In problem 5.3 we will ask the reader to derive a formula for $\text{Res}[g(z)/h(z),a]$ where $h(z)$ has a zero of order 2 at $z = a$.

If $f(z)$ has an ESP at $z = a$, we have no rule to offer for calculating $\text{Res}[f,a]$; in such a case $\text{Res}[f,a]$ is to be determined from the Laurent's series for $f(z)$ about $z = a$.

Example 5.3 Let $f(z) = e^z/z(z^2 + 1)$; $f(z)$ has simple poles at $z = 0$, i, $-i$.

In Theorem 5.2 let $g(z) = e^z$, $h(z) = z^3 + z$, and $h'(z) = 3z^2 + 1$. Then

$$\text{Res}[f,0] = \frac{g(0)}{h'(0)} = 1$$

$$\text{Res}[f,i] = \frac{g(i)}{h'(i)} = \frac{-e^i}{2}$$

$$\text{Res}[f,-i] = \frac{g(-i)}{h'(-i)} = \frac{-e^{-i}}{2}$$

Example 5.4 Let $f(z) = (\sin z)/z^3 = [(1/z^2)(\sin z/z)]$; by defining $(\sin z)/z$ to be 1 at $z = 0$, $(\sin z)/z$ is analytic at $z = 0$, and $f(z)$ has a pole of order 2 at $z = 0$. $\text{Res}[f,0] = \phi'(0)$, where

$$\phi(z) = \frac{\sin z}{z} = \sum_{n=0}^{\infty} \frac{(-1)^n z^{2n}}{(2n + 1)!},$$

and $\phi'(0) = 0$. Thus $\text{Res}[(\sin z)/(z^3),0] = 0$.

Example 5.5 Suppose C is the circle $\{z:|z + 1| = 2\}$ and $I = \int_C \{\exp(\pi z)/[z(z + 2)^3]\}\, dz$.

Since $f(z) = \exp(\pi z)/[z(z + 2)^3]$ has a simple pole at $z = 0$ and a pole of order 3 at $z = -2$,

$$I = 2\pi i\{\text{Res}[f,0] + \text{Res}[f,-2]\}.$$

$$\text{Res}[f,0] = \frac{\exp(\pi z)}{(z + 2)^3}\bigg|_{z=0} = \frac{1}{8}.$$

$$\text{Res}[f,-2] = \frac{\phi''(-2)}{2!},$$

where $\phi(z) = \exp(\pi z)/z$.

So $\dfrac{\phi''(-2)}{2!} = (-\tfrac{1}{8})(2\pi^2 + 2\pi - 1)\exp(-2\pi),$

and $I = \left(\dfrac{\pi i}{4}\right)[1 - (2\pi^2 + 2\pi - 1)\exp(-2\pi)].$

PROBLEMS

5.1   Find the residues for each function at its singularities.
   (a) $z + z^{-1}$
   (b) $z/(z^4 - 1)$
   (c)   csc z
   (d)   tan πz
   (e) $e^z/[(z - 1)(z - 2)(z - 3)]$
   (f) $\sin^2 z/z^3$
   (g) $z \exp(1/z)$
   (h) $e^z/[(z - 1)^3 \sin z]$

5.2   Evaluate each of the following integrals over the path indicated.

   (a) $\displaystyle\int_C \sin(1/2z)\, dz,\ C = \{z{:}|z| = 1\}$

   (b) $\displaystyle\int_C [\cos z/(z^2 + 1)]\, dz,\ C = \{z{:}|z| = 2\}$

   (c) $\displaystyle\int_C dz/(z^3 - 1),\ C = \{z{:}|z| = 2\}$

   (d) $\displaystyle\int_C (z + 2)/[z^2(z^2 - 1)]\, dz,\ C = \{z{:}|z + 1| = 3/2\}$

(e) $\displaystyle\int_C [\exp(1/z)/z]\, dz,\ C = \{z: |z| = 1\}$

(f) $\displaystyle\int_C [(\tan \pi z)/z]\, dz,\ C = \{z: |z| = 1\}$

5.3 Suppose $f(z) = g(z)/h(z)$, where $g(z)$ and $h(z)$ are analytic at $z = a$, $g(a) \neq 0$, and $h(z)$ has a zero of order 2 at $z = a$. Show that

$$\text{Res}[f,a] = \frac{6\, g'(a)\, h''(a) - 2\, g(a)\, h'''(a)}{3[h''(a)]^2}$$

5.4 Use problem 5.3 to help evaluate $\displaystyle\int_C z\, \csc^2 z\, dz$ where $C = \{z: |z| = 4\}$.

5.5 Suppose $f(z)$ is analytic at $z = a$ and has a zero of order m at $z = a$. If $f(z)$ has no other zeros on or inside $C = \{z: |z - a| = r\}$, show that

$$\frac{1}{2\pi i}\int_C \frac{f'(z)}{f(z)}\, dz = m.$$

5.6 If $f(z)$ is analytic on and inside a closed path C and has zeros at $a_1, a_2, \ldots, a_n$ inside C of orders $m_1, m_2, \ldots, m_n$ (but is otherwise different from zero on and within C), what is the value of

$$\frac{1}{2\pi i}\int_C \frac{f'(z)}{f(z)}\, dz?$$

5.7 Using problem 5.6, evaluate $(1/2\pi i)\displaystyle\int_C \tan \pi z\, dz$ for $C = \{z: |z| = 5\}$.

5.8 If $f(z)$ is analytic and has no zeros in the region $\{z: 0 < |z - a| \leq r\}$ and a pole of order n at $z = a$, show that

$$\frac{1}{2\pi i}\int_C \frac{f'(z)}{f(z)}\, dz = -n,$$

where $C = \{z: |z - a| = r\}$.

Section 5.2   Applications to Evaluation of Real Definite Integrals

Suppose $R(\sin \theta, \cos \theta)$ is a real-valued rational function of $\sin \theta$ and $\cos \theta$ for real values of θ ($R(\sin \theta, \cos \theta)$ is formed by adding, subtracting, multiplying, and dividing real multiples of integral powers of $\sin \theta$ and $\cos \theta$). In elementary calculus a technique was developed for evaluating $\int_0^{2\pi} R(\sin \theta, \cos \theta) \, d\theta$ which involves the following steps:

(1)   The substitution $y = \tan(\theta/2)$ transforms $R(\sin \theta, \cos \theta)$ to a rational function of y with real coefficients.

(2)   Each rational function of y with real coefficients has a partial fractions decomposition into pieces whose indefinite integrals are easily found.

(3)   By replacing each y with $\tan(\theta/2)$ in these indefinite integrals, we obtain an indefinite integral of $R(\sin \theta, \cos \theta)$.

If we let $z = \exp(i\theta) = \cos \theta + i \sin \theta$, then $\cos \theta = (\tfrac{1}{2})(z + z^{-1})$, $\sin \theta = (1/2i)(z - z^{-1})$, and (formally) $d\theta = dz/iz$. Integration in θ from 0 to 2π corresponds to integration in the positive direction about $\{z : |z| = 1\}$, so that $\int_0^{2\pi} R(\sin \theta, \cos \theta) \, d\theta$ can usually be evaluated by evaluating the integral of an appropriate rational function of z over the closed path $\{z : |z| = 1\}$.

Example 5.6   Suppose α is a real constant, $0 < \alpha < 1$. To evaluate $I(\alpha) = \int_0^{2\pi} d\theta/(1 + \alpha \cos \theta)$, we use the substitution $z = \exp(i\theta)$. Letting $C = \{z : |z| = 1\}$ we have

$$I(\alpha) = \int_C \frac{\dfrac{dz}{iz}}{1 + \dfrac{\alpha}{2}(z + z^{-1})} = \frac{2}{i} \int_C \frac{dz}{\alpha z^2 + 2z + \alpha}$$

Since $0 < \alpha < 1$, $0 < (1 - \alpha^2)^{1/2} < 1$, and $1/(\alpha z^2 + 2z + \alpha)$ has a simple pole at $[-1 + (1 - \alpha^2)^{1/2}]/\alpha$ as its only singularity on or inside C. Thus

$$I(\alpha) = \frac{2}{i}(2\pi i) \operatorname{Res}\left[\frac{1}{\alpha z^2 + 2z + \alpha}, \frac{(-1 + (1 - \alpha^2)^{1/2})}{\alpha} \right]$$

$$= 4\pi\left(\frac{1}{2(1 - \alpha^2)^{1/2}} \right) = \frac{2\pi}{(1 - \alpha^2)^{1/2}}.$$

In particular, we can see that

$$\int_0^{2\pi} \frac{d\theta}{\cos\theta + 2} = \frac{1}{2}\int_0^{2\pi} \frac{d\theta}{1 + \frac{1}{2}\cos\theta} = \frac{1}{2}\left[\frac{2\pi}{(1 - (\frac{1}{2})^2)^{1/2}}\right] = \frac{2\pi}{3^{1/2}}$$

The restriction on α in this example ensures that the original integral is proper. In cases where $\int_0^{2\pi} R(\sin\theta,\cos\theta)\, d\theta$ is improper — which will happen if the denominator of $R(\sin\theta,\cos\theta)$ vanishes for some value of θ — the transformation $z = \exp(i\theta)$ will produce an integral about $\{z:|z| = 1\}$ whose integrand has one or more poles on $\{z:|z| = 1\}$. In certain cases values can still be assigned to such improper integrals. We refer the reader to [Copson, pp. 133–135].

PROBLEMS

5.9 If $0 < \alpha < 1$, show that $\int_0^{2\pi} d\theta/(1 + \alpha\sin\theta) = 2\pi/(1 - \alpha^2)^{1/2}$, so that in fact $\int_0^{2\pi} d\theta/(1 + \alpha\cos\theta) = \int_0^{2\pi} d\theta/(1 + \alpha\sin\theta)$ for $0 < \alpha < 1$.

5.10 If $0 < \alpha < 1$, show that $\int_0^{2\pi} d\theta/(1 + \alpha\cos\theta)^2 = 2\pi/(1 - \alpha^2)^{3/2}$.

5.11 Use problem 5.10 to evaluate $\int_0^{2\pi} d\theta/(\beta + \cos\theta)^2$, for $\beta > 1$.

5.12 If n is a positive integer, show that

$$\int_0^\pi \sin^{2n}\theta\, d\theta = \frac{\pi(2n)!}{(2^n n!)^2}$$

Suppose $f(x)$ is a real-valued function of a real variable x which is defined for all real x. Let us recall two standard ways in which we agree to assign a meaning to the improper integral $\int_{-\infty}^\infty f(x)\, dx$.

(1) $\int_{-\infty}^\infty f(x)\, dx$ *converges* if for each real number α the following expression exists and has a value independent of α:

$$\lim_{R \to \infty} \int_\alpha^R f(x)\, dx + \lim_{S \to -\infty} \int_S^\alpha f(x)\, dx.$$

(2)   The *principal value* of $\int_{-\infty}^\infty f(x)\, dx$, written P.V. $\int_{-\infty}^\infty f(x)\, dx$,

exists if the limit $\lim_{r \to \infty} \int_{-r}^r f(x)\, dx$ exists, and

$$\text{P.V.} \int_{-\infty}^\infty f(x)\, dx = \lim_{r \to \infty} \int_{-r}^r f(x)\, dx.$$

We assume the reader is familiar with a variety of tests for deciding whether an improper integral either converges or has a principal value. A method involving integration by residues can help provide the values of many improper integrals whose convergence has already been established.

To outline the method, we take $f(x)$ to be real-valued and assume that $\int_{-\infty}^\infty f(x)\, dx$ converges. We try to find a complex-valued function, $F(z)$, of z and a closed path $C(R)$ containing a segment of the real axis, $-R \le x \le R$, so that

(a) $\int_{C(R)} F(z)\, dz$ can be evaluated by the residue theorem;

(b) $\lim_{R \to \infty} \int_{C(R)} F(z)\, dz$ either equals $\int_{-\infty}^\infty f(x)\, dx$ or is related to $\int_{-\infty}^\infty f(x)\, dx$ closely enough for us to evaluate $\int_{-\infty}^\infty f(x)\, dx$ from it.

Of course the power of the method rests on one's ingenuity in choosing $F(z)$ and $C(R)$ to accomplish (a) and (b).

Example 5.7   It is already known that $\int_{-\infty}^\infty dx/(x^2 + 4)$ converges to the value $\pi/2$. Let us verify this fact with residue techniques.

For $F(z)$ let us choose $F(z) = 1/(z^2 + 4)$; $F(z)$ is analytic in the finite plane except at $z = 2i, -2i$. Also $F(x) = 1/(x^2 + 4)$ for x real.

If $R > 2$, choose $C(R)$ to be the closed path consisting of the points $\{z: -R \le z \le R\}$ and the points $\{z: |z| = R, \text{Im } z \ge 0\}$. Now $F(z)$ is analytic on and inside $C(R)$ except for $z = 2i$, and

$$\int_{C(R)} F(z)\, dz = 2\pi i \, \text{Res}[F, 2i] = 2\pi i \left(\frac{1}{4i}\right) = \frac{\pi}{2}.$$

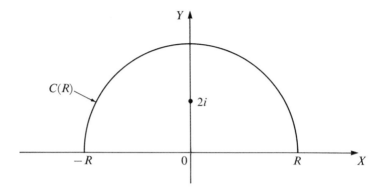

Figure 5.2

Note that

$$\int_{C(R)} F(z)\, dz = \frac{\pi}{2} \text{ for any } R > 2.$$

On the other hand,

$$\int_{C(R)} F(z)\, dz = \int_0^\pi F(Re^{i\theta})\, d(Re^{i\theta}) + \int_{-R}^R F(x)\, dx;$$

although the value of each integral on the right may change as R increases, their sum remains constant.

However,

$$\left| \int_0^\pi F(Re^{i\theta})\, d(Re^{i\theta}) \right| = \left| \int_0^\pi \frac{iRe^{i\theta}\, d\theta}{R^2 \exp(2i\theta) + 4} \right|$$

$$\leq \int_0^\pi \frac{R}{|R^2 \exp(2i\theta) + 4|}\, d\theta$$

$$\leq \int_0^\pi \left(\frac{R}{R^2 - 4} \right) d\theta = \frac{\pi R}{R^2 - 4},$$

as long as $R > 2$.

Thus

$$0 \le \lim_{R \to \infty} \left| \int_0^\pi F(Re^{i\theta}) \, d(Re^{i\theta}) \right| \le \lim_{R \to \infty} \left(\frac{\pi R}{R^2 - 4} \right) = 0,$$

or

$$\lim_{R \to \infty} \left| \int_0^\pi F(Re^{i\theta}) \, d(Re^{i\theta}) \right| = 0,$$

so that

$$\lim_{R \to \infty} \int_0^\pi F(Re^{i\theta}) \, d(Re^{i\theta}) = 0$$

as well.

Now we see that

$$\frac{\pi}{2} = \lim_{R \to \infty} \int_{C(R)} F(z) \, dz$$

$$= \lim_{R \to \infty} \int_0^\pi F(Re^{i\theta}) \, d(Re^{i\theta}) + \lim_{R \to \infty} \int_{-R}^R \frac{dx}{x^2 + 4},$$

$$\frac{\pi}{2} = 0 + \text{P.V.} \int_{-\infty}^\infty \frac{dx}{x^2 + 4} = \int_{-\infty}^\infty \frac{dx}{x^2 + 4}$$

since this last integral converges.

Without changing the details of this example substantially we can evaluate a considerable number of convergent improper integrals of the form $\int_{-\infty}^\infty f(x) \, dx$. In the hypotheses of the following theorem we state the conditions we want $f(x)$ to satisfy so that the argument of Example 5.7 will serve to evaluate $\int_{-\infty}^\infty f(x) \, dx$. The proof of the theorem is basically the argument of that example.

THEOREM 5.3 Hypotheses:

H1 $f(x)$ is a real-valued function for which P.V. $\int_{-\infty}^{\infty} f(x)\ dx$ exists.

H2 The related complex-valued function, $f(z)$, is such that:
    (a) $f(z)$ is analytic except for a finite number of singularities, none of which lies on the real axis;
    (b) $\lim_{R\to\infty} R|f(Re^{i\theta})| = 0$ for all θ, $0 \le \theta \le \pi$.

Conclusion: P.V. $\int_{-\infty}^{\infty} f(x)\ dx = 2\pi i \sum \text{Res}[f,a]$, where the sum is taken over all singularities of $f(z)$ *in the upper half-plane.*

(*Note:* If $\int_{-\infty}^{\infty} f(x)\ dx$ converges, then $\int_{-\infty}^{\infty} f(x)\ dx = \text{P.V.} \int_{-\infty}^{\infty} f(x)\ dx$, and the conclusion applies to $\int_{-\infty}^{\infty} f(x)\ dx$.)

Proof: Choose $R_0 > 0$ so large that all singularities of $f(z)$ occur at points z with $|z| < R_0$. For any $R > R_0$, define $C(R)$ as in Example 5.7 (Fig. 5.2).

By the residue theorem, $\int_{C(R)} f(z)\ dz = 2\pi i \sum \text{Res}[f,a]$, where this sum is taken over all singularities of $f(z)$ in the upper half-plane (since $R > R_0$). This sum is independent of R as long as $R > R_0$.

Also

$$\int_{C(R)} f(z)\ dz = \int_0^{\pi} f(Re^{i\theta})\ d(Re^{i\theta}) + \int_{-R}^{R} f(x)\ dx$$

$$= i \int_0^{\pi} Re^{i\theta} f(Re^{i\theta})\ d\theta + \int_{-R}^{R} f(x)\ dx.$$

Using hypothesis 2(b), given any number $\epsilon > 0$, we can find $R_1 > R_0$ so that $|Re^{i\theta} f(Re^{i\theta})| = R|f(Re^{i\theta})| < \epsilon/2\pi$ whenever $R > R_1$.

We know that

$$\lim_{R\to\infty} \int_{-R}^{R} f(x)\ dx = \text{P.V.} \int_{-\infty}^{\infty} f(x)\ dx$$

exists, so for some $R_2 > R_0$,

$$\left| \text{P.V.} \int_{-\infty}^{\infty} f(x)\, dx - \int_{-R}^{R} f(x)\, dx \right| < \frac{\epsilon}{2}$$

whenever $R > R_2$.

Now for any $R > \max(R_1, R_2)$,

$$\left| \text{P.V.} \int_{-\infty}^{\infty} f(x)\, dx - 2\pi i \sum \text{Res}[f, a] \right|$$

$$= \left| \text{P.V.} \int_{-\infty}^{\infty} f(x)\, dx - \int_{-R}^{R} f(x)\, dx - i \int_{0}^{\pi} Re^{i\theta} f(Re^{i\theta})\, d\theta \right|$$

$$\leq \left| \text{P.V.} \int_{-\infty}^{\infty} f(x)\, dx - \int_{-R}^{R} f(x)\, dx \right| + \int_{0}^{\pi} R|f(Re^{i\theta})|\, d\theta < \epsilon$$

Hence P.V. $\int_{-\infty}^{\infty} f(x)\, dx = 2\pi i \sum \text{Res}[f, a]$.

If hypothesis 2(b) is altered slightly, a similar result evaluates P.V. $\int_{-\infty}^{\infty} f(x)\, dx$ in terms of the residues of $f(z)$ at its singularities in the lower half-plane.

Example 5.8   The improper integral $\int_{0}^{\infty} [x^2/(x^4 + 1)]\, dx$ does converge. The function $f(z) = z^2/(z^4 + 1)$ satisfies the hypotheses of Theorem 5.3, and

$$\int_{0}^{\infty} \frac{x^2}{x^4 + 1}\, dx = \frac{1}{2} \int_{-\infty}^{\infty} \frac{x^2}{x^4 + 1}\, dx$$

$$= (\tfrac{1}{2})(2\pi i)\left\{ \text{Res}\left[f, \exp\left(\frac{\pi i}{4}\right) \right] + \text{Res}\left[f, \exp\left(\frac{3\pi i}{4}\right) \right] \right\}$$

$$= \pi i \left[\frac{z^2}{4z^3} \bigg|_{z=\exp(\pi i/4)} + \frac{z^2}{4z^3} \bigg|_{z=\exp(3\pi i/4)} \right]$$

$$= \frac{\pi}{2^{3/2}}.$$

Example 5.9 Let α be any nonzero real number and consider $I(\alpha) = \int_{-\infty}^{\infty} (\cos \alpha x)/(x^2 + 4) \, dx$, which is convergent. If we choose $f(z) = (\cos \alpha z)/(z^2 + 4)$ with an eye to using Theorem 5.3, then hypothesis 2(b) will not be satisfied because $|\cos \alpha z|$ is not bounded. However, $\cos \alpha z = \text{Re}[\exp(i\alpha z)]$, and $|\exp(i\alpha z)| = \exp(-\alpha y)$ is bounded in the upper half-plane if α is positive or the lower half-plane if α is negative.

Let us suppose that $\alpha > 0$ and let $f(z) = [\exp(i\alpha z)]/(z^2 + 4)$. For $R > 2$, take $C(R)$ as in Theorem 5.3, so that

$$\int_{C(R)} f(z) \, dz = 2\pi i \operatorname{Res}[f, 2i] = \frac{\pi \exp(-2\alpha)}{2},$$

while

$$\int_{C(R)} f(z) \, dz = \int_0^{\pi} \frac{\exp[i\alpha Re^{i\theta}]iRe^{i\theta}}{R^2 \exp(2i\theta) + 4} \, d\theta + \int_{-R}^{R} \frac{\exp(i\alpha x)}{x^2 + 4} \, dx$$

Now

$$\left| \int_0^{\pi} \frac{\exp[i\alpha Re^{i\theta}]iRe^{i\theta}}{R^2 \exp(2i\theta) + 4} \, d\theta \right| \leq \int_0^{\pi} \frac{R \exp(-\alpha \sin \theta)}{R^2 - 4} \, d\theta \leq \frac{R\pi}{(R^2 - 4)},$$

and

$$\lim_{R \to \infty} \left| \int_0^{\pi} \frac{\exp(i\alpha Re^{i\theta})}{R^2 \exp(2i\theta) + 4} \, d\theta \right| = 0.$$

Then we see that

$$\int_{-\infty}^{\infty} \frac{\exp(i\alpha x)}{x^2 + 4} \, dx = \frac{\pi \exp(-2\alpha)}{2},$$

and

$$\int_{-\infty}^{\infty} \frac{\cos \alpha x}{x^2 + 4} \, dx = \text{Re}\left[\int_{-\infty}^{\infty} \frac{\exp(i\alpha x)}{x^2 + 4} \, dx \right] = \frac{\pi \exp(-2\alpha)}{2},$$

if $\alpha > 0$.

In case $\alpha < 0$, we use a similar approach in the lower half-plane to obtain

$$\int_{-\infty}^{\infty} \frac{\cos \alpha x}{x^2 + 4} \, dx = \frac{\pi \exp(2\alpha)}{2}.$$

In general, $\int_{-\infty}^{\infty} (\cos \alpha x)/(x^2 + 4) \, dx = \pi \exp(-2|\alpha|)/2$ for any real α. (The case $\alpha = 0$ includes Example 5.7.)

The proof for the theorem below is very similar to the argument of this example. When the hypotheses of the theorem are satisfied we can evaluate convergent integrals of the form $\int_{-\infty}^{\infty} f(x)\cos \alpha x \, dx$, $\int_{-\infty}^{\infty} f(x)\sin \alpha x \, dx$ by examining the improper integral $\int_{-\infty}^{\infty} f(x)\exp(i\alpha x) \, dx$.

THEOREM 5.4   Hypotheses:
H1 $f(x)$ is a real-valued function for which the related complex-valued function $f(z)$
   (a)   is analytic except for a finite number of singularities, none of which lies on the real axis;
   (b)   satisfies $\lim_{R \to \infty} R|f(Re^{i\theta})| = 0$ for all θ.
H2 α is a real number.

Conclusions:
C1   If $\alpha \geq 0$,

$$\int_{-\infty}^{\infty} f(x)\exp(i\alpha x) \, dx = 2\pi i \sum \text{Res}[f(z)\exp(i\alpha z), a],$$

summing over the singularities of $f(z)$ in the *upper* half-plane.
C2   If $\alpha < 0$,

$$\int_{-\infty}^{\infty} f(x)\exp(i\alpha x) \, dx = -2\pi i \sum \text{Res}[f(z)\exp(i\alpha z), a],$$

summing over the singularities of $f(z)$ in the *lower* half-plane.

These results and examples do not give an accurate picture of the variety of closed paths or functions which may be used to evaluate convergent integrals of the form $\int_{-\infty}^{\infty} f(x)\, dx$. In the problems below we try to indicate additional approaches.

PROBLEMS

5.13 Use Theorems 5.3 and 5.4 to evaluate these improper integrals.

(a) $\int_{-\infty}^{\infty} dx/(x^2 + x + 1)$

(b) $\int_{0}^{\infty} dx/(x^2 + 4)^2$

(c) $\int_{-\infty}^{\infty} [(\cos 2x)/(x^2 + x + 1)]\, dx$

(d) $\int_{-\infty}^{\infty} [(x \sin \alpha x)/(x^4 + 1)]\, dx, \ \alpha > 0$

5.14 Evaluate $\int_{0}^{\infty} [(\cos^2 x)/(x^2 + 4)]\, dx$.

5.15 If m is a real number, evaluate $\int_{0}^{\infty} [(\sin mx)/x]\, dx$. (Let $f(z) = \exp(imz)/z$ for $m > 0$; for $0 < r < R$, let $C(r,R)$ be the closed path drawn in Fig. 5.3. Consider $\int_{C(r,R)} f(z)\, dz$ as $r \to 0, R \to \infty$.)

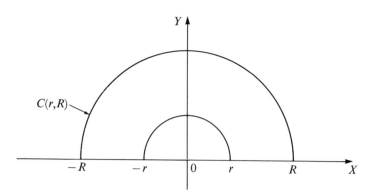

Figure 5.3

5.16 Show that $\int_0^\infty [(\sin^2 x)/x^2]\ dx = \pi/2$ using the closed path of problem 5.15.

5.17 Verify that for $0 < \alpha < 1$, $\int_{-\infty}^\infty [\exp(\alpha x)/(\exp x + 1)]\ dx = \pi/(\sin \pi\alpha)$. (For $R > 0$ let $C(R)$ be the boundary of the rectangle with vertices at $(R,0)$, $(R,2\pi)$, $(-R,2\pi)$, $(-R,0)$.)

5.18 Suppose p and q are positive integers with $p > q + 1$; then the improper integral $\int_0^\infty [x^q/(x^p + 1)]\ dx$ converges. To show that its value is $(\pi/p)\ \{\csc[\pi(\alpha + 1)/p]\}$, use $f(z) = z^q/(z^p + 1)$ and the closed path $C(R)$, $R > 1$, of Fig. 5.4.

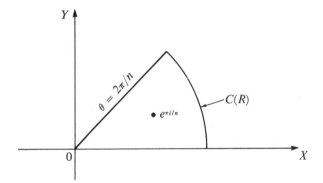

Figure 5.4

5.19 Show that

$$\int_0^\infty \frac{\sin x}{\sinh x}\ dx = \frac{\pi}{2}\left(\frac{1 + e^{-\pi}}{1 - e^{-\pi}}\right) = \frac{\pi}{2}\ \tanh\left(\frac{\pi}{2}\right)$$

by considering $f(z) = [\exp(iz)]/\sinh z$ and the closed path $C(r,R)$ of Fig. 5.5 for $0 < r < R$ as $r \to 0$, $R \to \infty$.

If the integrand, $f(x)$, in a convergent improper integral $\int_0^\infty f(x)\ dx$ is of a type such that the related complex-valued function, $f(z)$, is multiple-valued, additional care is necessary in both the choice of a closed path and the choice of a single-valued branch of $f(z)$. Once

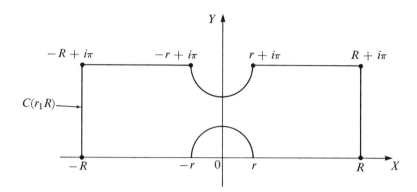

Figure 5.5

a single-valued branch of $f(z)$ is chosen, we choose a closed path which does not cross the branch cut but which still helps us evaluate

$$\int_0^\infty f(x)\,dx.$$

Example 5.10 Suppose $0 < \alpha < 1$ and we want to evaluate $\int_0^\infty [x^{\alpha-1}/(x+1)]\,dx$.

If we let $f(z) = z^{\alpha-1}/(z+1)$, we must choose a branch for $z^{\alpha-1}$. The principal branch for $z^{\alpha-1}$ has its branch cut along the negative real axis, and for this branch the behavior of $f(z)$ at $z = -1$ is quite involved. To keep the behavior at $z = -1$ simpler, let us choose a branch for $z^{\alpha-1}$ having branch cut along the positive real axis; let

$$z^{\alpha-1} = \exp[(\alpha-1)\log z] = \exp\{(\alpha-1)[\log|z| + i\arg z]\}$$

where $z \neq 0$ and $0 \leq \arg z < 2\pi$. Let $F(z) = z^{\alpha-1}/(z+1)$ with this branch of $z^{\alpha-1}$; $F(z)$ now has a simple pole at $z = -1$. With our closed path we want to enclose $z = -1$ but not cross the positive real axis. We use a "key-hole" type of path, $C(r,R,\epsilon)$ as in Fig. 5.6 with $0 < r < 1 < R$ and $0 < \epsilon < \pi/2$.
Now

$$\int_{C(r,R,\epsilon)} F(z)\,dz = 2\pi i\,\mathrm{Res}[F,-1] = 2\pi i(-1)^{\alpha-1} = 2\pi i\exp[(\alpha-1)i\pi],$$

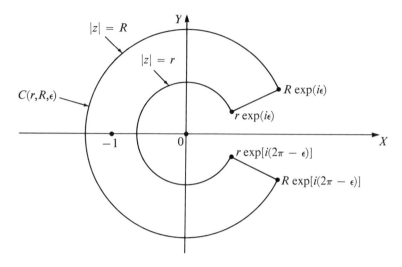

Figure 5.6

since $F(z)$ is analytic on and inside $C(r,R,\epsilon)$ except at $z = -1$. Note that this value will not change as R increases, r decreases, and ϵ decreases.
   Also

$$\int_{C(r,R,\epsilon)} F(z)\,dz = \int_r^R F(\rho e^{i\epsilon})\,d(\rho e^{i\epsilon}) + \int_\epsilon^{2\pi-\epsilon} F(Re^{i\theta})\,d(Re^{i\theta})$$

$$+ \int_R^r F\{\rho \exp[i(2\pi-\epsilon)]\}\,d\{\rho \exp[i(2\pi-\epsilon)]\}$$

$$+ \int_{2\pi-\epsilon}^\epsilon F(re^{i\theta})\,d(re^{i\theta})$$

$$= \int_r^R e^{i\epsilon} F(\rho e^{i\epsilon})\,d\rho$$

$$- \int_r^R \exp[i(2\pi-\epsilon)]\,F\{\rho\exp[i(2\pi-\epsilon)]\}d\rho +$$

$$i \int_{\epsilon}^{2\pi-\epsilon} R\, F(Re^{i\theta})e^{i\theta}\, d\theta$$

$$+\, i \int_{2\pi-\epsilon}^{\epsilon} r\, F(re^{i\theta})e^{i\theta}\, d\theta.$$

We can simplify the expressions somewhat, for

$$e^{i\epsilon}\, F(\rho e^{i\epsilon}) = \frac{e^{i\epsilon}\, \exp\{(\alpha-1)[\log \rho + i\epsilon]\}}{\rho e^{i\epsilon} + 1}$$

$$= \frac{e^{i\alpha\epsilon}\, \exp[(\alpha-1)\log \rho]}{\rho e^{i\epsilon} + 1} = \frac{e^{i\alpha\epsilon}\, \rho^{\alpha-1}}{\rho e^{i\epsilon} + 1},$$

and

$$\exp[i(2\pi-\epsilon)]F\{\rho \exp[i(2\pi-\epsilon)]\} = \frac{\exp[i\alpha(2\pi-\epsilon)]\rho^{\alpha-1}}{\rho e^{-i\epsilon} + 1}.$$

Then

$$2\pi i \exp[(\alpha-1)i\pi] = e^{i\alpha\epsilon}\int_{r}^{R} \frac{\rho^{\alpha-1}}{\rho e^{i\epsilon} + 1}\, d\rho - \exp[i\alpha(2\pi-\epsilon)]\int_{r}^{R} \frac{\rho^{\alpha-1}}{\rho e^{-i\epsilon} + 1}\, d\rho$$

$$+\, i \int_{\epsilon}^{2\pi-\epsilon} R\, F(Re^{i\theta})e^{i\theta}\, d\theta + i \int_{2\pi-\epsilon}^{\epsilon} r\, F(re^{i\theta})e^{i\theta}\, d\theta,$$

and as we let $\epsilon \to 0$,

$$2\pi i \exp[(\alpha-1)i\pi] = (1 - e^{i\alpha2\pi})\int_{r}^{R} \frac{\rho^{\alpha-1}}{\rho + 1}\, d\rho$$

$$+\, i \int_{0}^{2\pi} R\, F(Re^{i\theta})e^{i\theta}\, d\theta$$

$$+\, i \int_{2\pi}^{0} r\, F(re^{i\theta})e^{i\theta}\, d\theta.$$

However,

$$\left| i \int_0^{2\pi} R \, F(Re^{i\theta}) e^{i\theta} \, d\theta \right| \leq \int_0^{2\pi} R \left| \frac{\exp[(\alpha - 1)\log Re^{i\theta}]}{Re^{i\theta} + 1} \right| d\theta$$

$$\leq \int_0^{2\pi} \frac{R \, R^{\alpha-1}}{R - 1} \, d\theta = \frac{2\pi R^\alpha}{R - 1}.$$

Since $\alpha < 1$, $\lim_{R \to \infty} i \int_0^{2\pi} R \, F(Re^{i\theta}) e^{i\theta} \, d\theta = 0$.

Also,

$$\left| i \int_0^{2\pi} r \, F(re^{i\theta}) e^{i\theta} \, d\theta \right| \leq \int_0^{2\pi} r \left| \frac{\exp[(\alpha - 1)\log re^{i\theta}]}{re^{i\theta} + 1} \right| d\theta \leq \int_0^{2\pi} \frac{r \, r^{\alpha-1}}{1 - r} \, d\theta = \frac{r^\alpha}{1 - r}.$$

Since $0 < \alpha$, $\lim_{r \to 0} [r^\alpha/(1 - r)] = 0$, and

$$\lim_{r \to 0} i \int_0^{2\pi} r \, F(re^{i\theta}) e^{i\theta} \, d\theta = 0.$$

Taking limits now as $r \to 0$ and $R \to \infty$, we have

$$2\pi i \exp[(\alpha - 1)i\pi] = (1 - e^{i\alpha 2\pi}) \int_0^\infty \frac{\rho^{\alpha-1}}{\rho + 1} \, d\rho,$$

or

$$\int_0^\infty \frac{\rho^{\alpha-1}}{\rho + 1} \, d\rho = 2\pi i \left\{ \frac{\exp[(\alpha - 1)i\pi]}{1 - \exp(2\pi i \alpha)} \right\} = \frac{\pi}{\sin \pi \alpha}.$$

Thus

$$\int_0^\infty \frac{x^{\alpha-1}}{x + 1} \, dx = \frac{\pi}{\sin \pi \alpha} \quad \text{for} \quad 0 < \alpha < 1.$$

(There is an easier way to evaluate this particular integral. If we let $x = e^t$, the reader can show that

$$\int_0^\infty \frac{x^{\alpha-1}}{x+1}\, dx = \int_{-\infty}^\infty \frac{e^{\alpha t}}{e^t+1}\, dt,$$

and in problem 5.17 we found that this last integral has the value $\pi/\sin \pi\alpha$.)

PROBLEMS

5.20 Use integration by parts and Example 5.10 to show that
$\int_0^\infty [\log(1 + x)]/x^{1+\alpha}\, dx = \pi/(\alpha \sin \pi\alpha)$ for $0 < \alpha < 1$.

5.21 Use the closed path of Example 5.10 to show that

$$\int_0^\infty \frac{x^{1/2}}{(x^2 + 1)^2}\, dx = \frac{\pi}{2^{5/2}}.$$

5.22 To show $\int_0^\infty [(\log x)/(x^2 + 4)]\, dx = (\pi \log 2)/4$, use the principal branch of the logarithm and $f(z) = (\text{Log } z)^2/(z^2 + 4)$, together with the closed path $C(r,R,\epsilon)$ in Fig. 5.7 with $0 < r < 2 < R$ and $0 < \epsilon < \pi/2$.

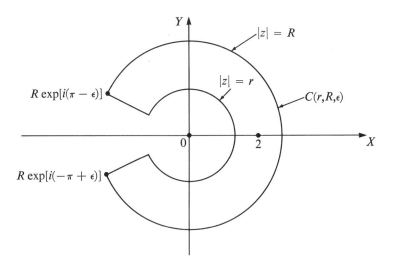

Figure 5.7

Let us use an example to illustrate one more technique hinted at in problem 5.19.

Example 5.11 The function $(\sin 2x)/x(x^4 - 16)$ is not defined for $x = 0, -2, 2$. The improper integral $\int_{-\infty}^{\infty} [(\sin 2x)/x(x^4 - 16)]\, dx$ does not converge, but we can establish the existence and value of P.V. $\int_{-\infty}^{\infty} [(\sin 2x)/x(x^4 - 16)]\, dx$.

Let us define $f(z) = \exp(2iz)/z(z^4 - 16)$; then $f(z)$ has simple poles at $z = 0, 2, -2, 2i, -2i$. For $0 < r < 1$ and $R > 3$ let $C(r,R)$ be the closed path of Fig. 5.8; we have "detoured" along semicircular paths to avoid the points $-2, 0, 2$ on the real axis.

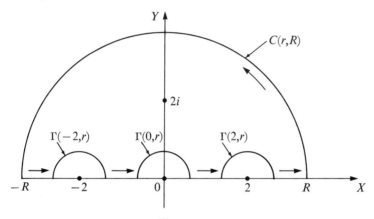

Figure 5.8

Now

$$\int_{C(r,R)} f(z)\, dz = 2\pi i \,\mathrm{Res}[f, 2i] = 2\pi i(e^{-4}/64) = \pi i e^{-4}/32.$$

If $\Gamma(a,r) = \{z : z = a + re^{i\theta}, 0 \le \theta \le \pi\}$, we can also write

$$\int_{C(r,R)} f(z)\, dz = \int_{0}^{\pi} f(Re^{i\theta})\, d(Re^{i\theta}) + \int_{-R}^{-2-r} f(x)\, dx$$

$$+ \int_{-2+r}^{0-r} f(x)\, dx + \int_{0+r}^{2-r} f(x)\, dx +$$

$$\int_{2+r}^{R} f(x)\, dx - \int_{\Gamma(-2,r)} f(z)\, dz$$

$$- \int_{\Gamma(0,r)} f(z)\, dz - \int_{\Gamma(2,r)} f(z)\, dz.$$

We may rearrange this sum of integrals to obtain

$$\left(\int_{-R}^{-2-r} + \int_{-2+r}^{0-r} + \int_{0+r}^{2-r} + \int_{2+r}^{R} \right)(f(x)\, dx)$$

$$= \frac{\pi i e^{-4}}{32} + \int_{\Gamma(-2,r)} f(z)\, dz + \int_{\Gamma(0,r)} f(z)\, dz$$

$$+ \int_{\Gamma(2,r)} f(z)\, dz - \int_{0}^{\pi} f(Re^{i\theta})\, d(Re^{i\theta}). \qquad (5.1)$$

We can see that

$$\left| \int_{0}^{\pi} f(Re^{i\theta})\, d(Re^{i\theta}) \right| \leq \int_{0}^{\pi} \frac{R\, |e^{2iz}|}{|z||z^4 - 16|}\, d\theta$$

$$\leq \int_{0}^{\pi} \frac{R \exp(-2R \sin \theta)}{R(R^4 - 16)}\, d\theta$$

$$\leq \frac{\pi}{(R^4 - 16)},$$

and we find that $\lim_{R \to \infty} \int_{0}^{\pi} f(Re^{i\theta})\, d(Re^{i\theta}) = 0$.

Next we examine $\int_{\Gamma(-2,r)} f(z)\, dz$ as $r \to 0$. Since $f(z)$ has a simple pole at $z = -2$, we write $f(z) = \mathrm{Res}[f,-2]/(z+2) + \phi(z)$, where $\phi(z)$ is analytic at $z = -2$, as long as z is sufficiently close to $z = -2$. Now $\phi(z)$ is continuous at $z = -2$, so let us choose r close enough to 0 so that $|\phi(z)| \leq 2|\phi(-2)|$ when $z = -2 + re^{i\theta}$.

Now for z on $\Gamma(-2,r)$,

$$\int_{\Gamma(-2,r)} f(z)\, dz = \int_{\Gamma(-2,r)} \frac{\mathrm{Res}[f,-2]}{z+2}\, dz + \int_{\Gamma(-2,r)} \phi(z)\, dz.$$

We see that

$$\int_{\Gamma(-2,r)} \frac{\operatorname{Res}[f,-2]}{z+2} \, dz = \int_0^\pi \frac{\operatorname{Res}[f,-2]}{re^{i\theta}} ire^{i\theta} \, d\theta = i\pi \operatorname{Res}[f,-2].$$

Also

$$\left| \int_{\Gamma(-2,r)} \phi(z) \, dz \right| \leq 2|\phi(-2)| \int_0^\pi r \, d\theta = 2\pi r |\phi(-2)|,$$

so that $\lim_{r \to 0} \int_{\Gamma(-2,r)} \phi(z) \, dz = 0$, and

$$\lim_{r \to 0} \int_{\Gamma(-2,r)} f(z) \, dz = i\pi \operatorname{Res}[f,-2] = i\pi e^{-4i}/64.$$

Similar arguments show that

$$\lim_{r \to 0} \int_{\Gamma(0,r)} f(z) \, dz = i\pi \operatorname{Res}[f,0] = \frac{-i\pi}{16},$$

and

$$\lim_{r \to 0} \int_{\Gamma(2,r)} f(z) \, dz = i\pi \operatorname{Res}[f,2] = \frac{i\pi e^{4i}}{64}.$$

In Equation (5.1), if we take limits as $r \to 0$ and $R \to \infty$, we find that

$$\text{P.V.} \int_{-\infty}^\infty f(x) \, dx = \frac{i\pi e^{-4}}{32} + \frac{i\pi e^{-4i}}{64} - \frac{i\pi}{16} + \frac{i\pi e^{4i}}{64}$$

$$= \frac{\pi i}{32} \left(e^{-4} - 2 + \frac{e^{4i} + e^{-4i}}{2} \right)$$

$$= \frac{\pi i}{32} (e^{-4} - 2 + \cos 4).$$

Since $(\sin 2x)/x(x^4 - 16) = \text{Im}\, f(x)$, it follows that

$$\text{P.V.} \int_{-\infty}^{\infty} \frac{\sin 2x}{x(x^4 - 16)}\, dx = \frac{\pi}{32}(e^{-4} - 2 + \cos 4).$$

PROBLEMS

5.23 Evaluate P.V. $\int_{-\infty}^{\infty} dx/(x^3 + 1)$ by the technique of Example 5.11.

5.24 Evaluate P.V. $\int_{0}^{\infty} (\cos x)/[(x^2 + 1)(x^2 - 4)]\, dx$ by the technique of Example 5.11.

Section 5.3 Counting the Zeros of an Analytic Function

Suppose $f(z)$ is analytic in a domain D except for a finite number of poles. The Residue Theorem provides us with an interesting relation between the orders of the zeros and poles of $f(z)$ and the residues of the function $f'(z)/f(z)$ at these points.

Let a be any point of D. If $f(z)$ is analytic at a and $f(a) \neq 0$, then $f'(z)/f(z)$ is analytic at $z = a$, and $\text{Res}[f'/f,a] = 0$. If $f(z)$ is analytic at $z = a$ and has a zero of order m there, problem 5.5 implies that $\text{Res}[f'/f,a] = m$. If $f(z)$ has a pole of order n at $z = a$, then $\text{Res}[f'/f,a] = -n$ (problem 5.8). That is, for $a \in D$,

$$\text{Res}[f'/f,a] = \begin{cases} 0 & \text{if} \quad a \text{ is neither a zero nor a pole of } f \\ m & \text{if} \quad f \text{ has a zero of order } m \text{ at } z = a \\ -n & \text{if} \quad f \text{ has a pole of order } n \text{ at } z = a \end{cases}$$

Using this information, together with an application of the Residue Theorem to $f'(z)/f(z)$, we obtain a theorem known as *the argument principle*, because of one of its geometric interpretations.

THEOREM 5.5 Hypotheses:
H1 $f(z)$ is analytic on a closed path C and has no zeros on C.
H2 inside C, $f(z)$ has zeros at $a_1, a_2, \ldots, a_j$ of orders $m_1, m_2, \ldots, m_j$.
H3 $f(z)$ is analytic inside C except for poles at $b_1, b_2, \ldots, b_k$ of orders $n_1, n_2, \ldots, n_k$.

Conclusion:

$$\frac{1}{2\pi i}\int_C \frac{f'(z)}{f(z)}\, dz = (m_1 + m_2 + \cdots + m_j) - (n_1 + n_2 + \cdots + n_k).$$

To rephrase the conclusion, if we count a zero (or pole) of order t as t zeros (or t poles), then

$$\frac{1}{2\pi i}\int_C \frac{f'(z)}{f(z)}\, dz$$

$= $ (Number of zeros of f inside C) $-$ (Number of poles of f inside C).

In particular, if f has no poles inside C — that is, $f(z)$ is analytic on and inside C — the integral $(1/2\pi i)\int_C f'(z)/f(z)\, dz$ counts the zeros of $f(z)$ inside C.

Outline of Proof: Choose a positive number δ so small that the circles $C(a_p,\delta) = \{z{:}|z - a_p| = \delta\}, p = 1, 2, \ldots, j$, and $C(b_q,\delta) = \{z{:}|z - b_q| = \delta\}$, $q = 1, 2, \ldots, k$, all lie inside C and no two of them intersect. As in the proof of the Residue Theorem, we see that

$$\frac{1}{2\pi i}\int_C \frac{f'(z)}{f(z)}\, dz = \sum_{p=1}^{j} \frac{1}{2\pi i}\int_{C(a_p,\delta)} \frac{f'(z)}{f(z)}\, dz$$

$$+ \sum_{q=1}^{k} \frac{1}{2\pi i}\int_{C(b_q,\delta)} \frac{f'(z)}{f(z)}\, dz$$

$$= \sum_{p=1}^{j} m_p - \sum_{q=1}^{k} n_q,$$

the last step being the application of problems 5.5 and 5.8.

If we consider what happens geometrically as one travels about C once in the positive direction, the connection between integration of f'/f and "zero-counting" may seem more plausible. Suppose $f(z)$ is analytic and never zero on C, and let $w = f(z)$. We start at some point z_1 on C, travel about C in the positive direction, and return to z_1. Meanwhile, in the w-plane, we start at $w_1 = f(z_1)$, trace out some curve, and return to w_1. Let us call this curve traced out in the w-plane Γ; Γ is closed but it may cross itself and so fail to be a path. Proceeding on intuitive grounds, we write

$$\frac{1}{2\pi i}\int_C \frac{f'(z)}{f(z)}\,dz = \frac{1}{2\pi i}\int_{\Gamma}^{w_1}\frac{dw}{w}$$

where we travel about Γ in whatever direction is induced by the positive direction about C. The latter integral is well-defined, since $f(z) \neq 0$ on C.
  Now

$$\frac{1}{2\pi i}\int_{\Gamma}^{w_1}\frac{dw}{w} = \frac{1}{2\pi i}[\log w]_{w_1}^{w_1},$$

where we must realize that if Γ has gone about $w = 0$ completely we may not be using the same branch of log w when we return to w_1 as we were when we left w_1. We must interpret $[1/(2\pi i)][\log w]_{w_1}^{w_1}$ as $1/(2\pi)$ times *the net change in* arg w as we have gone about Γ. And each time Γ goes about $w = 0$, arg w will increase or decrease by 2π as w starts from and returns to w_1. With $w = f(z)$ analytic and nonzero on C, we may interpret Theorem 5.5 in this way:

[Number of zeros of $f(z)$ inside C]

$$= \frac{1}{2\pi i}\int_C \frac{f'(z)}{f(z)}\,dz$$

$= \frac{1}{2\pi}$[change in arg $f(z)$ as z goes about C once in the positive direction]

$= $ [Number of times $\Gamma = \{w = f(z) : z \in C\}$ winds about $w = 0$ as z goes about C once in the positive direction].

  With Theorem 5.5 we can derive a result which helps us count the zeros inside a closed path for an analytic function by comparing it with another analytic function whose distribution of zeros is already known. The result is known as *Rouché's Theorem*.

THEOREM 5.6   Hypotheses:
H1 $f(z)$ and $g(z)$ are analytic on and inside a closed path C.
H2   On C, $|g(z)| < |f(z)|$.

Conclusion: $f(z)$ and $f(z) + g(z)$ have the same number of zeros inside C.

Proof: H2 implies neither $f(z)$ nor $f(z) + g(z)$ can be zero at any point on C. If M_C is the number of zeros of $f(z) + g(z)$ inside C, and N_C is the number of zeros of $f(z)$ inside C, we may use Theorem 5.5 to write

$$M_C - N_C = \frac{1}{2\pi i} \int_C \left(\frac{f'(z) + g'(z)}{f(z) + g(z)} \right) dz - \frac{1}{2\pi i} \int_C \frac{f'(z)}{f(z)} dz$$

$$= \frac{1}{2\pi i} \int_C \left\{ \frac{f(z)g'(z) - f'(z)g(z)}{[f(z)]^2 + f(z)g(z)} \right\} dz$$

$$= \frac{1}{2\pi i} \int_C \left\{ \frac{\left[\dfrac{g(z)}{f(z)} \right]'}{1 + \left[\dfrac{g(z)}{f(z)} \right]} \right\} dz$$

$$= \frac{1}{2\pi i} \int_C \frac{F'(z)}{F(z)} dz, \quad \text{where } F(z) = \frac{g(z)}{f(z)} + 1$$

and $f(z) \neq 0$ on C.

At points on C, $|F(z) - 1| = |g(z)/f(z)| < 1$; this implies that the closed curve described by $w = F(z)$ as z goes about C cannot enclose the point $w = 0$. Thus

$$\frac{1}{2\pi i} \int_C \frac{F'(z)}{F(z)} dz = 0, \quad \text{and} \quad M_C = N_C.$$

Example 5.12   Let $p(z) = a_0 + a_1 z + \cdots + a_n z^n$ be any polynomial of degree $n \geq 1$, with $a_n \neq 0$. In problem 3.20 we used Liouville's Theorem to show that for at least one point z_0, $p(z_0) = 0$. With some elementary algebra it is possible to show that the equation $p(z) = 0$ has exactly n solutions if multiple roots are counted according to their multiplicity. This fact is also obtainable from Rouché's Theorem.

Let $C(R)$ be the circle $\{z : |z| = R\}$ for $R > 1$. Write $f(z) = a_n z^n$ and $g(z) = a_0 + a_1 z + \cdots + a_{n-1} z^{n-1}$, so that $p(z) = f(z) + g(z)$.

On $C(R)$, $|f(z)| = |a_n| R^n$ while

$$|g(z)| = |a_0 + a_1 z + \cdots + a_{n-1} z^{n-1}|$$

$$\leq |a_0| + |a_1| R + \cdots + |a_{n-1}| R^{n-1}$$

$$\leq (|a_0| + |a_1| + \cdots + |a_{n-1}|) R^{n-1}$$

as long as $R > 1$.

If we choose $R_0 \geq (|a_0| + |a_1| + \cdots + |a_{n-1}|)/|a_n|$, when z is on $C(R)$ for any $R > R_0$,

$$|g(z)| \leq (|a_0| + |a_1| + \cdots + |a_{n-1}|)R^{n-1} < (|a_n|R)R^{n-1} = |f(z)|.$$

By Rouché's Theorem, $f(z) = a_n z^n$ and $p(z) = f(z) + g(z)$ have the same number of zeros inside $C(R)$. Since $f(z)$ has a zero of order n at $z = 0$ and no other zeros, $p(z)$ has n zeros inside $C(R)$ for any $R > (|a_0| + |a_1| + \cdots + |a_{n-1}|)/|a_n|$.

Example 5.13 If α is a constant, $\alpha > e$, let us show that the equation $e^z - \alpha z^n = 0$ has n solutions in $\{z : |z| < 1\}$.

Let $f(z) = -\alpha z^n$ and $g(z) = e^z$. For all z, $|z| = 1$, $|g(z)| = \exp[Re\, z] \leq e$, while $|f(z)| = \alpha|z|^n = \alpha > e$. By Rouche's Theorem, αz^n and $f(z) + g(z) = e^z - \alpha z^n$ have the same number of zeros inside $\{z : |z| = 1\}$. Since αz^n has a zero of order n at $z = 0$ and no other zeros, $e^z - \alpha z^n = 0$ must have n solutions inside $\{z : |z| = 1\}$.

PROBLEMS

5.25 Let $p(z) = 2z^4 + 3z^3 - z^2 + z + 4$. Find a number $\beta > 0$ so that you can be sure that all the zeros of $p(z)$ lie inside $\{z : |z| = \beta\}$.

5.26 Show that $5z^n = \cosh 2z$ has n solutions in the region $\{z : |z| < 1\}$.

5.27 Show that in $\{z : |z| < 1\}$ the equation $3z^4 - \sin z = 0$ has four solutions. One of them obviously is $z = 0$. Of the remaining solutions, how many are real?

5.28 Use Theorem 5.5 to show that e^z never has value 0.

5.29 Suppose $f(z)$ is analytic for $|z - a| \leq r$, $f(z)$ has a zero of order m at $z = a$, but $f(z) \neq 0$ for $0 < |z - a| \leq r$. Show that, if $C = \{z : |z - a| = r\}$,

$$\frac{1}{2\pi i} \int_C \frac{z f'(z)}{f(z)}\, dz = ma.$$

5.30 Suppose $f(z)$ is analytic and never zero for $0 < |z - a| \leq r$ but has a pole of order n at $z = a$. Show that, if $C = \{z : |z - a| = r\}$,

$$\frac{1}{2\pi i} \int_C \frac{z f'(z)}{f(z)}\, dz = -na.$$

5.31 Suppose $f(z)$ is analytic on and inside a closed path C, $f(z) \neq 0$ on C, except for poles at a_1, a_2, a_3 inside C of orders p_1, p_2, p_3. If $f(z)$ has zeros at b_1, b_2, b_3 inside C of orders q_1, q_2, q_3, what value should we expect

$$\frac{1}{2\pi i} \int \frac{z f'(z)}{f(z)} \, dz \text{ to have?}$$

5.32 Use your answer to the previous problem to evaluate:

(a) $\int_C [(2z^2 + z)/(z^2 + z + 1)] \, dz$ for $C = \{z : |z| = 2\}$

(b) $\int_C z \tan \pi z \, dz$ for $C = \{z : |z - 1| = 1\}$.

5.33 Suppose $f(z)$ is analytic for $|z| \leq r$, $f(0) = 0$, $f'(0) \neq 0$, and $f(z) \neq 0$ for $0 < |z| \leq r$. Let $\rho = \min_{|z|=r} |f(z)|$. Use Rouché's Theorem to show that for each point w, $|w| < \rho$, there is exactly one solution in $\{z : |z| < r\}$ for the equation $f(z) = w$. (Note that the hypotheses imply that $f(z)$ has a simple zero at $z = 0$ as its only zero on or inside the circle $\{z : |z| = r\}$.)

5.34 In problem 5.33 let $z = g(w)$ be the solution to the equation $f(z) = w$ in $\{z : |z| < r\}$ (that is, $f[g(w)] = w$).

As w takes on values $|w| < \rho$, $z = g(w)$ determines points in $\{z : |z| < r\}$. Use problem 5.29 with the function $F(t) = f(t) - w$ to show that

$$z = g(w) = \frac{1}{2\pi i} \int_C \frac{t f'(t)}{f(t) - w} \, dt$$

where C is the circle $\{t : |t| = r\}$.

The point of the last two problems has been to show that if $w = f(z)$ is analytic in $\{z : |z| \leq r\}$ with $f(0) = 0$, $f'(0) \neq 0$, we can find a number $\rho > 0$ so that for $|w| < \rho$ we can solve the equation $f(z) = w$ uniquely for z. This means that an inverse function $z = g(w)$ exists for $|w| < \rho$ with the property that $f[g(w)] = w$. The integral form for $g(w)$ in problem 5.34 makes it fairly easy to show that $g(w)$ is analytic for $|w| < \rho$. Theorem 1.6 will imply that $g'(0) = 1/f'[g(0)] = 1/f'(0)$.

We can make a similar statement if $f(z_0) = w_0$ and $f'(z_0) \neq 0$, and we can summarize these remarks in a theorem which will be useful in Chapter 6.

THEOREM 5.7 Hypotheses: $w = f(z)$ is analytic at $z = z_0$, $w_0 = f(z_0)$, and $f'(z_0) \neq 0$.

Conclusions: There exist numbers $r > 0$ and $\rho > 0$ such that:
C1 for each w, $|w - w_0| < \rho$, the equation $f(z) = w$ has exactly one
     solution $z = g(w)$ in $\{z : |z - z_0| < r\}$;
C2 the function $z = g(w)$ defined in $\{w : |w - w_0| < \rho\}$ is analytic there;
C3 $g'(w_0) = 1/f'(z_0)$.

Section 5.4   Fourier Transforms

Integral transforms of various types are very helpful in treating many differential equation problems of interest to physicists and engineers. To illustrate the utility of an integral transform technique, and to provide an additional application of the residue theory of Section 5.1, we shall look briefly at the Fourier transform.

We do not attempt to motivate the definition of the Fourier transform or to prove its most important properties. For a more complete discussion we refer the reader to [Carrier, Chapter 7].

If $f(x)$ is a real-valued function defined for all x, and y is a real variable, we define the *Fourier transform of $f(x)$* to be

$$F(y) = \frac{1}{(2\pi)^{1/2}} \int_{-\infty}^{\infty} f(x)e^{-iyx}\, dx.$$

Of course for $F(y)$ to be defined for any values of y we must require $f(x)$ to satisfy some conditions guaranteeing the existence of this improper integral. In addition, we are considering the Fourier transform as a problem-solving tool, so we would like $F(y)$ to exist and behave nicely enough for us to be able to "recover" $f(x)$ from $F(y)$. Consequently, we define a class of functions, to be used throughout our discussion, which will have well-behaved Fourier transforms.

Definition   We shall say that a real-valued function $f(x)$, defined for all real x, is in the *class D* if:
     (1) $f(x)$ is continuous for all but a finite number of values of x;

(2)   at each x_0 where $f(x)$ is not continuous, both these one-sided limits exist:

$$\lim_{x \to x_0 (x > x_0)} f(x) = f(x_0^+),$$

$$\lim_{x \to x_0 (x < x_0)} f(x) = f(x_0^-);$$

(3) $f(x)$ is absolutely integrable, meaning that the improper integral, $\int_{-\infty}^{\infty} |f(x)|\, dx$, converges.

If we restrict our attention to functions in class D, we can state this important theorem.

THEOREM 5.8   Hypothesis: $f(x)$ is in the class D.

Conclusions:

C1   The Fourier transform of f, $F(y) = [1/(2\pi)^{1/2}] \int_{-\infty}^{\infty} f(x)e^{-iyx}\, dx$, is defined for all real y.

C2   For each real x

$$\frac{1}{(2\pi)^{1/2}} \int_{-\infty}^{\infty} F(y)e^{ixy}\, dy = \frac{f(x^+) + f(x^-)}{2}.$$

C2 allows us to recover from $F(y)$ the values of f where it is continuous, and the average of the left and right-sided limits of f where it is not continuous. In C2 we are "inverting" the Fourier transform $F(y)$, and we call the integral in C2 the *inverse Fourier transform* of $F(y)$.

Since the Fourier transform is a mapping of functions in class D into a class of functions possessing inverse Fourier transforms, we can think of the transform as a "function" whose domain and range are themselves collections of functions. If $f(x)$ is in class D, we frequently let $\mathcal{F}(f)$ denote the Fourier transform, $F(y)$, of f and $\mathcal{F}^{-1}(F)$ denote the inverse Fourier transform of F.

Any function $f(x)$ which satisfies the hypotheses of Theorem 5.4 is in the class D, and so we can use Theorem 5.4 to find the Fourier transform in many cases.

Example 5.14   Let $f(x) = 1/(x^2 + 1)$; clearly $f(x)$ is in class D and satisfies the hypotheses of Theorem 5.4. Then

$$F(y) = \frac{1}{(2\pi)^{1/2}} \int_{-\infty}^{\infty} \frac{e^{-iyx}}{x^2 + 1} \, dx$$

$$= \begin{cases} i(2\pi)^{1/2} \, \mathrm{Res}\!\left[\dfrac{e^{-iyx}}{(x^2 + 1)}, \, i\right], \, y \leq 0 \\[2mm] -i(2\pi)^{1/2} \, \mathrm{Res}\!\left[\dfrac{e^{-iyx}}{(x^2 + 1)}, \, -i\right], \, y > 0 \end{cases}$$

$$= \left(\frac{\pi}{2}\right)^{1/2} \exp(-|y|).$$

C2 claims that

$$\frac{1}{(2\pi)^{1/2}} \int_{-\infty}^{\infty} \left(\frac{\pi}{2}\right)^{1/2} e^{-|y|} e^{ixy} \, dy = \frac{1}{x^2 + 1};$$

let us verify that this is the case.

$$\frac{1}{(2\pi)^{1/2}} \int_{-\infty}^{\infty} \left(\frac{\pi}{2}\right)^{1/2} e^{-|y|} e^{ixy} \, dy$$

$$= \left(\frac{1}{2}\right)\left\{ \int_{-\infty}^{0} \exp[y(1 + ix)] \, dy + \int_{0}^{\infty} \exp[y(-1 + ix)] \, dy \right\}$$

$$= \frac{1}{2}\left(\frac{1}{1 + ix} - \frac{1}{-1 + ix}\right) = \frac{1}{x^2 + 1}.$$

Properties of the Fourier Transform

By listing a few basic properties of Fourier transforms we can begin to see why they are so helpful in solving some kinds of differential equations.

(1) If $f_1(x)$ and $f_2(x)$ are in class D, then for any constants α and β, $\alpha f_1(x) + \beta f_2(x)$ is in class D, and

$$\mathcal{F}(\alpha f_1 + \beta f_2) = \alpha \mathcal{F}(f_1) + \beta \mathcal{F}(f_2).$$

(This property follows directly from the linear properties of integration.)

(2) If $f(x)$ is in class D and $\mathcal{F}(f) = F(y)$, while a and b are real numbers with $a \neq 0$, then $g(x) = f(ax + b)$ is in class D, and

$$\mathfrak{F}(g) = e^{iby}\frac{F\left(\dfrac{y}{a}\right)}{|a|}.$$

That is, $\mathfrak{F}[f(ax + b)] = e^{iby} F(y/a)/|a|$. To see why this property is at least formally correct, we write

$$\mathfrak{F}(g) = \frac{1}{(2\pi)^{1/2}} \int_{-\infty}^{\infty} f(ax + b)e^{-iyx}\, dx$$

and use the change of variables $u = ax + b$ to obtain

$$\mathfrak{F}(g) = \frac{1}{(2\pi)^{1/2}} \int_{-\infty}^{\infty} f(u)\exp\left[\frac{-iy(u - b)}{a}\right]\frac{du}{|a|}$$

$$= \frac{e^{iyb}}{(2\pi)^{1/2}|a|} \int_{-\infty}^{\infty} f(u)\exp\left(\frac{-iyu}{a}\right) du$$

$$= \frac{e^{iyb}}{|a|}F\left(\frac{y}{a}\right).$$

(3) If $f(x)$ is in class D and has k derivatives, $f'(x), f''(x), \ldots,$ $f^{(k)}(x)$, also in class D, then for $j = 1, 2, 3, \ldots, k,$

$$\mathfrak{F}(f^{(j)}) = (iy)^j \mathfrak{F}(f).$$

Let us examine this statement for $j = 1$. If $f'(x)$ is in class D, then $\mathfrak{F}(f')$ exists, and

$$\mathfrak{F}(f') = \frac{1}{(2\pi)^{1/2}} \lim_{R \to \infty} \int_{-R}^{R} f'(x)e^{-iyx}\, dx.$$

If we integrate by parts,

$$\int_{-R}^{R} f'(x)e^{-iyx}\, dx = f(x)e^{-iyx}\big]_{-R}^{R} + iy \int_{-R}^{R} f(x)e^{-iyx}\, dx.$$

Since $|e^{-iyx}| = 1$ and $f(x)$ is in class D, the contributions at $x = R$ and $x = -R$ approach 0 as $R \to \infty$. Then if we take limits as $R \to \infty$, we see that $\mathfrak{F}(f') = iy\,\mathfrak{F}(f)$.

For higher order derivatives we repeat this process of integration by parts.

(4)   If $f(x)$ is in class D, let $h(x) = \int_a^x f(t)\, dt$ for a fixed real number a. If $h(x)$ is in class D, then

$$\mathfrak{F}(h) = \left(\frac{1}{iy}\right)\mathfrak{F}(f).$$

If $h(x)$ is in class D, since $h'(x) = f(x)$, we can use property (3) to see that $(iy)\mathfrak{F}(h) = \mathfrak{F}(h') = \mathfrak{F}(f)$.

The last property we want to use unfortunately does not have a simple motivation. If $f(x)$ and $g(x)$ are in class D, it is often very helpful to be able to identify what function in class D, if any, has the Fourier transform $\mathfrak{F}(f)\mathfrak{F}(g)$. Formally we can provide a function, a type of integral product of f and g, whose Fourier transform is $\mathfrak{F}(f)\mathfrak{F}(g)$. This function, called the *convolution* of f and g, is defined to be

$$h(x) = (f * g)(x) = \frac{1}{(2\pi)^{1/2}} \int_{-\infty}^{\infty} f(t)g(x - t)\, dt$$

if this integral indeed defines a function. We cannot claim that the invention of $(f * g)(x)$ is obvious or natural, but it is effective and helpful. However, for some ordinary differential equations over real intervals of finite length, a convolution does provide the solution; in this setting a convolution can be thought of as natural. (See problem 5.40.)

To state property (5) we shall assume that the needed convolutions exist and also have Fourier transforms.

(5)   If $f(x)$ and $g(x)$ are in class D, and if the convolution

$$(f * g)(x) = \frac{1}{(2\pi)^{1/2}} \int_{-\infty}^{\infty} f(t)g(x - t)\, dt$$

is a well-defined function of x with a Fourier transform, then so is

$$(g * f)(x) = \frac{1}{(2\pi)^{1/2}} \int_{-\infty}^{\infty} g(t)f(x - t)\, dt,$$

and $(f * g)(x) = (g * f)(x)$,

$$\mathfrak{F}(f * g) = \mathfrak{F}(g * f) = \mathfrak{F}(f)\mathfrak{F}(g).$$

For a formal justification of this last equation, we have

$$\mathfrak{F}(f * g) = \frac{1}{(2\pi)^{1/2}} \int_{-\infty}^{\infty} \left[\frac{1}{(2\pi)^{1/2}} \int_{-\infty}^{\infty} f(t)g(x - t)\, dt \right] e^{-iyx}\, dx$$

$$= \frac{1}{(2\pi)} \int_{-\infty}^{\infty} \int_{-\infty}^{\infty} f(t)g(x - t)e^{-iyx}\, dt\, dx$$

$$= \frac{1}{(2\pi)} \int_{-\infty}^{\infty} \int_{-\infty}^{\infty} f(t)g(s)\exp[-iy(s + t)]\, dt\, ds$$

$$= \left[\frac{1}{(2\pi)^{1/2}} \int_{-\infty}^{\infty} f(t)e^{-iyt}\, dt \right]\left[\frac{1}{(2\pi)^{1/2}} \int_{-\infty}^{\infty} g(s)e^{-iys}\, ds \right]$$

$$= \mathfrak{F}(f)\mathfrak{F}(g).$$

PROBLEMS

5.35 Verify that each function below is in class D and find its Fourier transform.
   (a) $f(x) = 1/(x^2 + 4)$
   (b) $f(x) = 1/(x^2 + 1)^2$
   (c) $f(x) = 1/(x^2 + x + 1)$
   (d) $f(x) = \begin{cases} e^{-ax}, & x \geq 0 \\ 0, & x < 0 \end{cases}$, a some positive constant
   (e) $f(x) = \begin{cases} 1 - |x|, & |x| \leq 1 \\ 0, & |x| > 1 \end{cases}$

5.36 In each part of problem 5.35 decide whether $\mathfrak{F}^{-1}[\mathfrak{F}(f)] = f(x)$; if not, what will $\mathfrak{F}^{-1}[\mathfrak{F}(f)]$ be?

5.37 Let $f(x) = \begin{cases} k, & |x| \leq 1 \\ 0, & |x| > 1 \end{cases}$ where k is a positive constant.
   (a) Find $\mathfrak{F}(f)$.
   (b) Find $(f * f)(x)$.
   (c) Verify that $\mathfrak{F}(f * f) = [\mathfrak{F}(f)]^2$.
   (d) Is the inverse transform of $[\mathfrak{F}(f)]^2$ equal to $f * f$?

5.38 Find the Fourier transform of the following functions:
   (a) $f(x) = \begin{cases} 1, & 0 \le x \le 1 \\ 0, & x < 0, x > 1 \end{cases}$
   (b) $f(x) = \exp(-|x|)$
   (c) $f(x) = \exp(-|x|) \cos 2x$

5.39 Suppose $f(x)$ and $[f(x)]^2$ are in class D.
   (a) If $g(x) = f(-x)$, verify that $\mathfrak{F}(g) = \overline{\mathfrak{F}(f)}$.
   (b) Use $(f * g)(0)$ to show that

$$\int_{-\infty}^{\infty} [f(x)]^2 \, dx = \int_{-\infty}^{\infty} [\mathfrak{F}(f)]^2 \, dy.$$

   (This relation is known as *Parseval's identity* for Fourier transforms.)

5.40 Suppose $f(x)$ is continuous on the interval $0 \le x \le \pi$ and k is a positive constant.
   (a) Show that the only solutions to the problem

$$\left\{ \begin{aligned} y''(x) + k^2 y(x) &= 0, \, 0 < x < \pi \\ y(0) &= 0 \end{aligned} \right\}$$

   are constant multiples of $\sin kx$.
   (b) Verify that the related problem

$$\left\{ \begin{aligned} y''(x) + k^2 y(x) &= f(x), \, 0 < x < \pi \\ y(0) &= 0 \end{aligned} \right\}$$

   has $y(x) = (1/k) \int_0^x f(t) \sin k(x - t) \, dt$ as a solution.

5.41 Given that $\int_{-\infty}^{\infty} \exp(-x^2) \, dx = (\pi)^{1/2}$, verify that $f(x) = \exp(-x^2/2)$ has $F(y) = \exp(-y^2/2)$ as its Fourier transform.

Applications to Differential Equations

Suppose $f(x)$ is a solution to a certain n^{th}-order linear ordinary differential equation with constant coefficients:

$$a_n f^{(n)}(x) + a_{n-1} f^{(n-1)}(x) + \cdots + a_1 f'(x) + a_0 f(x) = g(x).$$

Assuming that $f(x)$ and $g(x)$ have Fourier transforms, we can use the properties of Fourier transforms to obtain

$$[a_n(iy)^n + a_{n-1}(iy)^{n-1} + \cdots + a_1 iy + a_0]\mathfrak{F}(f) = \mathfrak{F}(g)$$

or
$$P(iy)\mathfrak{F}(f) = \mathfrak{F}(g)$$

where
$$P(z) = a_n z^n + a_{n-1}z^{n-1} + \cdots + a_1 z + a_0.$$

Then $\mathfrak{F}(f) = \mathfrak{F}(g)/P(iy)$, and if we can find a function $h(x)$ such that $\mathfrak{F}(h) = 1/P(iy)$,

$$\mathfrak{F}(f) = \mathfrak{F}(g)\mathfrak{F}(h).$$

Thus we obtain by formal means a possible solution to the original equation:

$$f(x) = \frac{1}{(2\pi)^{1/2}} \int_{-\infty}^{\infty} g(t)h(x - t)\, dt.$$

This candidate must be tested in the original equation.

Example 5.15 Let A, B, C be positive constants with $A \neq B$. Let us use the method of Fourier transforms on the equation $A f'(x) + B f(x) = C \exp(-|x|)$.

Transforming the equation, we have (by problem 5.38(b))

$$A\, iy\, \mathfrak{F}(f) + B\, \mathfrak{F}(f) = \left(\frac{2}{\pi}\right)^{1/2}(1 + y^2)^{-1},$$

and
$$\mathfrak{F}(f) = \left(\frac{2}{\pi}\right)^{1/2}[(iyA + B)(1 + y^2)]^{-1}.$$

If we invert this transform, we obtain

$$f(x) = \frac{e^{-x}}{B - A} + \frac{2 A \exp\left(\dfrac{-xB}{A}\right)}{A^2 - B^2}, \quad x > 0,$$

and
$$f(x) = \frac{e^x}{B + A}, \quad x \leq 0$$

which actually does solve the differential equation.

In several types of partial differential equation problems Fourier transforms may be used to reduce a problem to an ordinary differential equation problem which might be solved by one of several methods. In these cases also, any "solution" produced by formal manipulations of Fourier transforms must be tested to see if it solves the actual problem.

As an illustration, let us consider the problem of obtaining a solution $u(x,t)$ to the partial differential equation problem:

$$\begin{cases} u_t - k^2\, u_{xx} = 0, & t > 0, \ -\infty < x < \infty \\ u(x,0) \quad\quad = f(x), & -\infty < x < \infty \end{cases}$$

With appropriate physical assumptions, this problem serves as a model for finding the temperature, $u(x,t)$, at each time t at a point x on a long thin wire, when we know that initially the temperature at each point x was given by $f(x)$. The partial differential equation is called the heat equation in one dimension.

We assume that for each fixed time t our solution $u(x,t)$ has a Fourier transform and that $f(x)$ has a Fourier transform. We hold t fixed and transform in the x variable; if we let

$$U(t) = \frac{1}{(2\pi)^{1/2}} \int_{-\infty}^{\infty} u(x,t)e^{-iyx}\, dx,$$

formally we write

$$U'(t) = \frac{1}{(2\pi)^{1/2}} \int_{-\infty}^{\infty} u_t(x,t)e^{-iyx}\, dx,$$

$$-y^2 U(t) = \frac{1}{(2\pi)^{1/2}} \int_{-\infty}^{\infty} u_{xx}(x,t)e^{-iyx}\, dx,$$

and our problem becomes

$$\begin{cases} U'(t) + k^2 y^2\, U(t) = 0, & t > 0 \\ U(0) = \mathfrak{F}(f) & \end{cases}$$

This is a standard initial-value problem for ordinary differential equations, whose unique solution for each fixed y is

$$U(t) = \mathfrak{F}(f)\exp(-k^2 y^2 t), \quad t \geq 0.$$

Inverting the transform, we want $u(x,t)$ to be

$$u(x,t) = \frac{1}{(2\pi)^{1/2}} \int_{-\infty}^{\infty} \mathfrak{F}(f)\exp(-k^2y^2t)\exp(ixy) \, dy.$$

Or if $h(x,t)$ is a function such that $\mathfrak{F}(h) = \exp(-k^2y^2t)$ for each fixed t, we want

$$u(x,t) = (f * h)(x,t)$$

$$= \frac{1}{(2\pi)^{1/2}} \int_{-\infty}^{\infty} f(s)h(x - s,t) \, ds,$$

where the convolution is formed for each fixed t.

PROBLEMS

5.42 Use the method of Fourier transforms to look for a solution to the ordinary differential equation

$$f''(x) - f(x) = \exp(-|x|).$$

Does what you obtain really solve the differential equation?

5.43 If k is a positive constant and for each $t \geq 0$ $\mathfrak{F}h(x,t) = \exp(-k^2y^2t)$, find $h(x,t)$.

5.44 If $f(x)$ and $g(x)$ are functions in class D, k is a positive constant, and $u(x,t)$ is the solution to the problem

$$\begin{cases} u_{tt} - k^2 u_{xx} = 0, & t > 0, \ -\infty < x < \infty \\ u(x,0) = f(x), & -\infty < x < \infty \\ u_t(x,0) = g(x), & -\infty < x < \infty \end{cases}$$

let $U(t,y) = [1/(2\pi)^{1/2}] \int_{-\infty}^{\infty} u(x,t)e^{-iyx} \, dx$ for each $t \geq 0$.

(a) Verify that formally $U(t,y)$ should satisfy the following equations for each fixed y:

$$\frac{d^2U(t,y)}{dt^2} + k^2y^2U(t,y) = 0, \quad t > 0$$

$$U(0,y) = \mathcal{F}(f)$$

$$\left. \frac{dU(t,y)}{dt} \right|_{t=0} = \mathcal{F}(g)$$

(b)   Solve this ordinary differential equation problem for $U(t,y)$.

(c)   Determine the inverse transform of $U(t,y)$ in the special case where
      $f(x) = 1/(x^2 + 1)$ and $g(x) = 1/(x^2 + 1)$.

6

Conformal Mapping

In Chapter 1 we studied complex-valued functions of a complex variable by looking at pairs of real-valued functions of two real variables, and we considered the function as a mapping of the complex plane into the complex plane. If $f(z)$ is a function defined in some set D of the complex z-plane, we can more easily study the behavior of $f(z)$ as a mapping by visualizing the image $w = f(z)$ of each point of D as a point in another complex plane, the w-plane.

The study of the mapping properties of analytic functions of a complex variable is interesting both in its applications and in its mathematical content. In this chapter we present a brief sample of both of these aspects.

Section 6.1 Mapping by Analytic Functions

Suppose $w = f(z)$ is analytic in a domain D in the z-plane, and Δ is the range of f on D. In the remaining sections of this chapter we shall want $f(z)$ to be 1-1 on D and to have a nonzero derivative at each point of D; in this section we try to explain why we want $f(z)$ to have all of these properties.

If $f(z)$ is 1-1 on D, then for each point w of Δ there is exactly one point z of D with $f(z) = w$, and we can define a function $g(w)$ on Δ by

$$z = g(w) \quad \text{for} \quad w \in \Delta, \quad \text{where} \quad f[g(w)] = w$$

Therefore, associated with $w = f(z)$ we have a 1-1 inverse mapping $z = g(w)$ from Δ to D.

If $f'(z) \neq 0$ for each z in D, we can apply Theorem 5.7. For any $w_0 \in \Delta$ there is a number $\rho > 0$ such that $\{w : |w - w_0| < \rho\} \subset \Delta$, so that Δ is an open set; the inverse mapping $z = g(w)$ from Δ to D is analytic in Δ and $g'(w) = 1/f'[g(w)]$ for each $w \in \Delta$.

Finally, for any points w_1, w_2 in Δ let $z_1 = g(w_1)$ and $z_2 = g(w_2)$. Since z_1, z_2 are in D, a domain, there is a path C in D from z_1 to z_2 which $w = f(z)$ will map onto a path in Δ from w_1 to w_2; since we noted earlier that Δ is open, it follows that Δ is a domain.

Let us summarize these remarks in a formal statement.

THEOREM 6.1 Hypotheses:
H1 D is a domain.
H2 $f(z)$ is analytic in D with range Δ.
H3 For each $z \in D, f'(z) \neq 0$.
H4 $w = f(z)$ is a 1-1 mapping of D onto Δ.

Conclusions:
C1 Δ is a domain.
C2 The function $g(w)$ defined on Δ by $f[g(w)] = w$ is analytic in Δ and $g'(w) = 1/f'[g(w)] \neq 0$.
C3 $z = g(w)$ is a 1-1 mapping of Δ onto D.

Now it is apparent why we wish to consider mappings of a domain D by a 1-1 analytic function with nonvanishing derivative: The image of D under such a mapping is also a domain Δ, and the inverse mapping from Δ to D is also accomplished by a 1-1 analytic function with nonvanishing derivative.

Example 6.1 Let $D_1 = \{z = re^{i\theta} : 1 < r < 2, 0 < \theta < \pi\}$, $D_2 = \{z = re^{i\theta} : 1 < r < 2, 0 < \theta < 2\pi\}$, and $w = f(z) = z^2$.

We see that $f(z)$ is analytic and has a non-vanishing derivative at each point of D_1 and D_2. The mapping $w = z^2$ maps D_1 1-1 onto $\Delta_1 = \{w = \rho e^{i\phi} : 1 < \rho < 4, 0 < \phi < 2\pi\}$; it also maps D_2 onto $\Delta_2 = \{w = \rho e^{i\phi} : 1 < \rho < 4, 0 < \phi < 4\pi\}$ but the mapping is not 1-1.

Thus the mapping $w = z^2$ from D_1 onto Δ_1 has an analytic inverse mapping defined by

$$z = \rho^{1/2} \exp\!\left(\frac{i\phi}{2}\right) \quad \text{for} \quad \rho e^{i\phi} \text{ in } \Delta_1$$

which maps Δ_1 1-1 onto D_1.

The mapping $w = z^2$ from D_2 onto Δ_2 does not have this property, since for every point $w_0 \in \Delta_2$ there are two distinct points, z_{01} and z_{02}, of D_2 for which $(z_{01})^2 = w_0 = (z_{02})^2$. However, $f(z) = z^2$ has a nonzero derivative throughout D_2. This means that we can find a neighborhood $N(w_0)$ about w_0 and neighborhoods $N(z_{01})$ and $N(z_{02})$ about z_{01} and z_{02}, respectively, such that $w = z^2$ does have an analytic inverse function which maps $N(w_0)$ 1-1 onto $N(z_{01})$ and an analytic inverse function which maps $N(w_0)$ 1-1 onto $N(z_{02})$. Thus C2 and C3 of Theorem 6.1 apply when we consider the mapping $w = z^2$ from D_1 onto Δ_1, but they do not apply when we consider the mapping $w = z^2$ from D_2 onto Δ_2. In the latter case H4 is not satisfied.

Suppose $w = f(z)$ is analytic in a domain D containing $z = z_0$, and let $w_0 = f(z_0)$. Let C be any path in D passing through z_0 and possessing a tangent line at z_0. We let α denote the inclination of the tangent to C at z_0, measured with respect to the positive real axis. (See Fig. 6.1.)

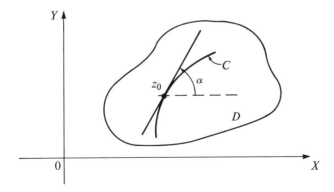

Figure 6.1

If we write $f(z)$ in a Taylor's series about z_0,

$$w = f(z) = f(z_0) + \sum_{k=1}^{\infty} \frac{f^{(k)}(z_0)}{k!}(z - z_0)^k,$$

we have

$$w - w_0 = f(z) - f(z_0) = \sum_{k=1}^{\infty} \frac{f^{(k)}(z_0)}{k!}(z - z_0)^k.$$

If j is the smallest positive integer for which $f^{(j)}(z_0) \neq 0$,

$$w - w_0 = (z - z_0)^j \sum_{k=j}^{\infty} \frac{f^{(k)}(z_0)}{k!}(z - z_0)^{k-j} = (z - z_0)^j F(z),$$

where $F(z)$ is analytic at z_0 and $F(z_0) = f^{(j)}(z_0)/j!$. Also $\arg(w - w_0) = j \operatorname{Arg}(z - z_0) + \operatorname{Arg} F(z)$.

Now $w = f(z)$ maps the path C onto a curve Γ in the w-plane passing through $w = w_0$ and having a tangent line at w_0. As z approaches z_0 along C, w approaches w_0 along Γ, $\arg(z - z_0) \to \alpha$, and $\arg F(z) \to \arg F(z_0) = \arg f^{(j)}(z_0)$, this last quantity being a number independent of the particular curve C along which we approach z_0. Then

$$\lim_{z \to z_0 (z \in C)} \arg(w - w_0) = j \alpha + \arg f^{(j)}(z_0).$$

Thus the inclination of the tangent to Γ at w_0 is given by $j \alpha + \arg f^{(j)}(z_0)$.

Now, as in Fig. 6.2, we look at paths C_1 and C_2 in D intersecting at z_0 with tangent lines at z_0 of inclinations α_1 and α_2, respectively. In the

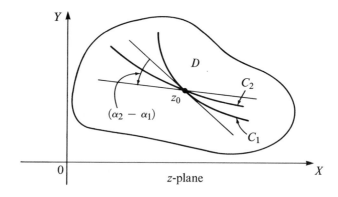

Figure 6.2

w-plane the corresponding curves Γ_1 and Γ_2 through w_0 have tangent lines at w_0 with inclinations $[\arg f^{(j)}(z_0) + j\,\alpha_1]$ and $[\arg f^{(j)}(z_0) + j\,\alpha_2]$, respectively, as drawn in Fig. 6.3.

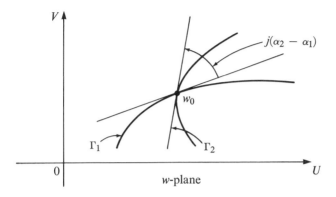

Figure 6.3

If we measure the angle, $\measuredangle(C_1,C_2,z_0)$, from C_1 to C_2 at z_0 by the angle from the tangent to C_1 to the tangent to C_2 in the positive direction and likewise the angle, $\measuredangle(\Gamma_1,\Gamma_2,w_0)$, from Γ_1 to Γ_2 at w_0, we see that $\measuredangle(C_1,C_2,z_0) = \alpha_2 - \alpha_1$ and $\measuredangle(\Gamma_1,\Gamma_2,w_0) = j(\alpha_2 - \alpha_1)$.

Example 6.2   If $f(z) = z^2$ and $z_0 = 0$, $f'(0) = 0$, $f''(0) = 2$. Then if C_1 and C_2 are two paths intersecting at $z = 0$ with $\measuredangle(C_1,C_2,0) = \gamma$, the mapping $w = z^2$ will yield two curves Γ_1 and Γ_2 intersecting at $w = 0$ with $\measuredangle(\Gamma_1,\Gamma_2,0) = 2\gamma$.

In the foregoing discussion it is clear that if $f'(z_0) \neq 0$, then $\measuredangle(C_1,C_2,z_0) = \measuredangle(\Gamma_1,\Gamma_2,w_0)$, and the mapping $w = f(z)$ "preserves angles" at z_0. This is another important reason for wanting to consider mapping by functions with nonzero derivatives. At each point where a mapping $w = f(z)$ has this (geometric) angle-preserving property we say the mapping is *conformal*. We have indicated that if $w = f(z)$ is analytic in a domain D and $f'(z_0) \neq 0$ at each point of D, then the mapping $w = f(z)$ is conformal at each point of D. As a partial converse, if $f(z)$ is a function defined in a domain D and the real and imaginary parts of $f(z)$ have continuous first partial derivatives with respect to x and y at each point of D, the assumption that the mapping $w = f(z)$ is conformal in D implies that $f(z)$ is analytic and $f'(z) \neq 0$ at each point of D. A fuller discussion and some details of proof for these remarks may be found in [Hille, pp. 95–97].

Example 6.3 Suppose $w = f(z) = u(x,y) + i \ v(x,y)$ is analytic at $z_0 = x_0 + i \ y_0$, $w_0 = f(z_0) = \alpha + i\beta$, and $f'(z_0) \neq 0$. In a small neighborhood of w_0 in the w-plane the perpendicular lines $u = \alpha$, $v = \beta$ intersecting at w_0 are mapped by the inverse function $z = g(w)$ of $w = f(z)$ onto curves $u(x,y) = \alpha$, $v(x,y) = \beta$ in the z-plane intersecting at z_0. Since $f'(z_0) \neq 0$, $g(w)$ is analytic at w_0, $g'(w_0) = 1/f'(z_0) \neq 0$, and the inverse mapping is conformal at w_0. This means that the curves $u(x,y) = \alpha$, $v(x,y) = \beta$ intersect at right angles at z_0. This is the result of problem 1.31, and in the language used there we say that the level curves, $u(x,y) = \alpha$ and $v(x,y) = \beta$ of the conjugate pair of harmonic functions, $u(x,y)$ and $v(x,y)$, are orthogonal at their point of intersection z_0. Of course, $w = f(z)$ maps the lines $x = x_0$ and $y = y_0$ onto curves in the w-plane which intersect and are orthogonal at w_0.

If $f(z)$ is analytic in a domain D and has a nonzero derivative at each point of D, it may yet be difficult to verify that the mapping $w = f(z)$ is 1-1 on D. We present one useful criterion for this property in the theorem below; for a nice collection of criteria for determining when a mapping is 1-1 we refer the reader to [Kaplan, pp. 124–125].

THEOREM 6.2 Hypotheses:

H1 $f(z)$ is analytic on a closed path C and in the domain D enclosed by C.

H2   The image of C under $w = f(z)$ is a closed path Γ; and as z goes about C once in the positive sense, $w = f(z)$ goes about Γ once in the positive sense.

Conclusion: $w = f(z)$ maps D 1-1 onto the domain Δ enclosed by Γ.

The proof for this theorem is outlined in problem 6.6. It is worth noting that Theorem 6.2 is still valid if we replace H1 by this less restrictive hypothesis: $f(z)$ is continuous on a closed path C and its interior D, and is analytic in D.

Behind the applications of conformal mappings in mathematical and physical problems lie two important theorems we wish to state without proof. We can, however, comment on them and point out our use of them in later sections.

The first theorem gives conditions under which a function, 1-1 and analytic on a simply-connected domain D can be extended to the boundary of D so as to be continuous on the union of D and its boundary. It is this theorem which makes valid the solutions to "boundary-value" problems, as we will illustrate later.

THEOREM 6.3   Hypotheses:
H1 D is a simply-connected domain whose boundary C is a closed path.
H2 $f(z)$ is analytic in D, $f'(z) \neq 0$ in D.
H3 $w = f(z)$ maps D 1-1 onto a simply-connected domain Δ in the
w-plane whose boundary is a closed path Γ.

Conclusions: It is possible to define $w = f(z)$ at each point of C in such
a way that:
C1 $w = f(z)$ maps C 1-1 onto Γ;
C2 $f(z)$ is continuous on $D \cup C$.

The second theorem is known as the *Riemann mapping theorem.*

THEOREM 6.4   Hypothesis: D is any simply-connected domain in
the z-plane other than the entire complex plane.

Conclusion: There exists an analytic function, $f(z)$, which maps D 1-1
and conformally onto the open unit disk $\{w : |w| < 1\}$.

Let D_1 and D_2 be two simply-connected domains, neither of which is
the entire plane. Then there must be functions $f_1(z)$ and $f_2(z)$, analytic in
D_1 and D_2 respectively, which map D_1 and D_2 1-1 and conformally onto
$\{w : |w| < 1\}$. If $z = g_2(w)$ is the inverse mapping of $w = f_2(z)$ from
$\{w : |w| < 1\}$ onto D_2, then $g_2(w)$ is analytic, 1-1, and conformal on
$\{w : |w| < 1\}$. Now form the function $F(z) = g_2[f_1(z)]$ for z in D_1. It is

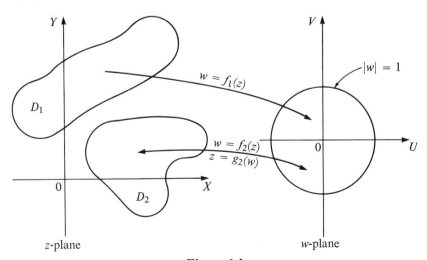

Figure 6.4

easy to see that $F(z)$ is analytic in D_1, has images in D_2, and will determine a 1-1 conformal mapping of D_1 onto D_2. We say two simply connected domains, neither of which is the entire plane, are *conformally equivalent*, meaning that there is a 1-1 conformal mapping from one to the other. Each such domain is conformally equivalent to the open unit disk. The open unit disk, then, will serve as a completely representative model of such domains whenever we deal with properties or phenomena which are not altered by conformal mappings — for example, angles between curves at their points of intersection.

However, the Riemann mapping theorem is only an "existence" theorem; it asserts only that a mapping with certain properties does exist. There will be in general many such mappings, and finding or constructing any one of them may be a difficult problem.

Example 6.4 Let $D = \{z : \text{Im } z > 0\}$ be the upper half plane. We will verify that the mapping $w = (z - i)/(z + i)$ is a 1-1 conformal mapping of D onto $\{w : |w| < 1\}$.

If $z = x + iy$ is any point of D, then $y > 0$ and

$$|w|^2 = \left|\frac{x + i(y - 1)}{x + i(y + 1)}\right|^2 = \frac{x^2 + (y - 1)^2}{x^2 + (y + 1)^2} = \frac{x^2 + y^2 - 2y + 1}{x^2 + y^2 + 2y + 1} < 1$$

Then for $z \in D$, $|(z - i)/(z + i)| < 1$. (At the same time we can see that if z is in the lower half plane, then $|(z - i)/(z + i)| > 1$.)

For any w_0, where $|w_0| < 1$, the equation $w_0 = (z - i)/(z + i)$ has a unique solution $z = i(1 + w_0)/(1 - w_0)$ in D, so the mapping is 1-1 in D.

The reader can verify that $(z - i)/(z + i)$ is differentiable with nonzero derivative at each point of D.

The mapping $w = (z - i)/(z + i)$ carries $z = i$ to $w = 0$, the center of the unit disk. Theorem 6.3 does not apply directly to the problem of mapping $C = \{z : |z| = 1\}$ onto $\Gamma = \{w : |w| = 1\}$. It is not possible to give the details here to show that we can extend Theorem 6.3 to the unbounded domains we will consider in this chapter whose boundaries are paths, but we shall assume that this is true. In the example above, the mapping $w = (z - i)/(z + i)$ will map C 1-1 onto Γ. In fact, there are many functions analytic in D which will map D 1-1 and conformally onto $\{w : |w| < 1\}$, and all of them are, or can be, defined on C, mapping C 1-1 onto Γ. (See problem 6.4.) The unanswered question in this example is: How does one choose for himself a mapping to map D onto $\{w : |w| < 1\}$? This question is quite different from verifying that a given mapping has certain properties.

PROBLEMS

6.1   Let $D = \{z: -1 < \text{Re } z < 1, -\pi/2 < \text{Im } z < \pi/2\}$ and C be the boundary of D. If Δ is the image of D under the mapping $w = e^z$ and Γ is its boundary:
(a)   sketch Δ and Γ;
(b)   use Theorem 6.2 to show that e^z is 1-1 on D.

6.2   Let $D = \{z: 0 < \text{Re } z < 1, 0 < \text{Im } z < 1\}$ and C be its boundary. Find and sketch the image of D and C under the mapping $w = \sin z$.

6.3   If τ and α are constants, $\tau > 0, 0 \le \alpha < 2\pi$, use some branch of the logarithm function as a mapping to prove that the circle $\{z: |z| = \tau\}$ and the ray $\arg z = \alpha$ intersect orthogonally.

6.4   If w_0 is any point with $|w_0| < 1$, and $z = i[(1 + w_0)/(1 - w_0)]$, show that $\text{Im } z > 0$.

6.5   Suppose $D = \{z: \text{Im } z > 0\}$ and α is any fixed point of D. Let $f(z) = (z - \alpha)/(z - \bar{\alpha})$ for $z \in D$.
(a)   Show that $w = f(z)$ maps D 1-1 and conformally onto $\Delta = \{w: |w| < 1\}$.
(b)   Show that $w = f(z)$ maps $C = \{z: \text{Im } z = 0\}$ 1-1 onto $\Gamma = \{w: |w| = 1\}$.

6.6   Assume that the hypotheses of Theorem 6.2 are satisfied. The results (a), (b), (c), (d) will prove Theorem 6.2.

(a)   Use the argument principle to show that

$$\frac{1}{2\pi i}\int_C \frac{f'(z)}{f(z) - w_0}\, dz = \begin{cases} 1, & \text{if } w_0 \text{ is inside } \Gamma \\ 0, & \text{if } w_0 \text{ is outside } \Gamma \end{cases}$$

This means that for each w_0 inside Γ there is only one z_0 inside C with $f(z_0) = w_0$; and for any w_0 outside Γ there is no z_0 inside C with $f(z_0) = w_0$.

(b)   Suppose for some point $w_1 \in \Gamma$ there exists a point $z_1 \in D$ with $f(z_1) = w_1$. Use problem 4.27(a) to show that for some $\delta > 0$ with $\{z: 0 < |z - z_1| < \delta\} \subset D$, $f(z) \ne w_1$ if $0 < |z - z_1| < \delta$.

(c)   Now let $G = \{z: |z - z_1| = \delta/2\}$. Use part (b), the argument principle, and the fact that w_1 lies on Γ to show that $f(G) = \{w = f(z): z \in G\}$ contains at least one point outside Γ.

(d)   Conclude from (a) and (c) that there is no point in D where $f(z)$ has a value on Γ.

Section 6.2   Some Conformal Mappings

In lieu of a more complete and general treatment of mappings by analytic functions, we shall illustrate the Riemann mapping theorem by looking at a variety of simply-connected domains and finding analytic functions which map them 1-1 and conformally onto either the open unit disk or the upper half plane. (The last two regions are conformally equivalent in the light of Example 6.4 and problem 6.5.) In each case Theorem 6.3 will apply to ensure a 1-1 correspondence between the boundary of the domain and the boundary of the range of each function.

By taking a case-by-case approach we merely illustrate some standard mapping problems which, together with a list of problems, will acquaint the reader with some techniques of mapping by elementary functions. Several more complete catalogs of standard conformal mappings of the type we illustrate are available; for example, [Kober]; [Churchill, pp. 284–291]; [Spiegel, pp. 205–211].

Open Disk onto Open Unit Disk

If β is any real number and α is any point of the open unit disk $\{z:|z| < 1\}$, then $w = e^{i\beta}[(z - \alpha)/(1 - \bar{\alpha}z)]$ is an analytic function which maps $\{z:|z| < 1\}$ 1-1 and conformally onto $\{w:|w| < 1\}$.

Proof: If $|\alpha| < 1$ and $|z| < 1$, then $(1 - |\alpha|^2)(1 - |z|^2) > 0$, and $1 + |\alpha|^2|z|^2 > |\alpha|^2 + |z|^2$. But then

$$|w|^2 = \frac{|z - \alpha|^2}{|1 - z\bar{\alpha}|^2} = \frac{|\alpha|^2 + |z|^2 - 2\,\mathrm{Re}(\bar{\alpha}z)}{1 + |\alpha|^2|z|^2 - 2\,\mathrm{Re}(\bar{\alpha}z)} < 1,$$

and $|w| < 1$ whenever $|z| < 1$. The reader can verify that the mapping is 1-1 on $\{z:|z| < 1\}$. The conformality follows from the fact that for

$$f(z) = e^{i\beta}\left(\frac{z - \alpha}{1 - \bar{\alpha}z}\right), \quad f'(z) = \frac{1 - \alpha}{(1 - \bar{\alpha}z)^2} \neq 0$$

in $\{z:|z| < 1\}$.

Now if $D = \{z:|z - a| = r\}$ is any open disk, $\xi = g(z) = (z - a)/r$ clearly maps D 1-1 and conformally onto $\{\xi:|\xi| < 1\}$, while $f(\xi) = e^{i\beta}[(\xi - \alpha)/(1 - \bar{\alpha}\xi)]$, for any real β and $|\alpha| < 1$, maps $\{\xi:|\xi| < 1\}$ 1-1 and conformally onto $\{w:|w| < 1\}$. Therefore $w = f[g(z)] =$

$e^{i\beta}\{\tau(z - \gamma)/[\tau^2 - (\overline{\gamma - a})(z - a)]\}$ maps D 1-1 and conformally onto $\{w:|w| < 1\}$ with $\alpha = (\gamma - a)/\tau$ and $f[g(\gamma)] = 0$.

Example 6.5 If γ is any point of $D = \{z:|z - 3i| < 2\}$, for any real β, $w = e^{i\beta}\{2(z - \gamma)/[4 - (\overline{\gamma - 3i})(z - 3i)]\}$ will map D 1-1 and conformally onto $\{w:|w| < 1\}$ with $z = \gamma$ corresponding to $w = 0$. If we specify the image of a point on the boundary of D, a value of β, $-\pi \le \beta < \pi$, will be determined.

Suppose that we ask $\gamma = 2i$ to be mapped to $w = 0$ and $z = 5i$ to be mapped to $w = 1$. Then $1 = 6ie^{i\beta}/[4 - i(2i)] = ie^{i\beta}$, and $e^{i\beta} = -i$, $\beta = -\pi/2$.

Exterior of Closed Disk onto Open Unit Disk

A mapping which is frequently useful in applied mathematics is one which maps the domain $\{z:|z| > 1\}$ 1-1 and conformally onto $\{w:|w| < 1\}$. We shall assert that the mapping $w = 1/z$ does exactly this, even though we shall not fully justify that this is the case. It is easy to see that $f(z) = 1/z$ is analytic with nonzero derivative for each z, $|z| > 1$, and that for each point w, $0 < |w| < 1$, there is exactly one value of z such that $w = 1/z$.

For numbers $\tau > 1$, whenever $|z| > \tau$, $|w| = 1/|z| < 1/\tau$, so that $w = 1/z$ maps points z with large modulus to points close to $w = 0$. To provide a point in the z-plane which corresponds to $w = 0$ under this mapping requires us to invent a "point at infinity" — which might be described as the point which lies outside $|z| = \tau$ for every $\tau > 0$ — and then to show that it is possible to define the derivative of $f(z) = 1/z$ at such a point and to argue that the mapping $w = 1/z$ is indeed 1-1 and conformal at and in the neighborhood of this point. (Those who wish to pursue this may look at [Pennisi, pp. 61–63, 305–309].)

Example 6.6 Let us find an analytic function to map the region $D = \{z:|z - 2| > \frac{1}{2}\}$ 1-1 and conformally onto the region $\Delta = \{w:\text{Im } w > 0\}$.

First, $\xi_1 = (z - 2)/(\frac{1}{2}) = 2(z - 2)$ maps D 1-1 and conformally onto $D_1 = \{\xi_1:|\xi_1| > 1\}$. Next $\xi_2 = 1/\xi_1$ maps D_1 1-1 and conformally onto $D_2 = \{\xi_2:|\xi_2| < 1\}$. Finally $w = i[(1 + \xi_2)/(1 - \xi_2)]$ maps D_2 1-1 and conformally onto Δ. If we compose all these mappings, we obtain a 1-1 conformal mapping of D onto Δ given by:

$$z \rightarrow \xi_1 = \frac{z - 2}{\frac{1}{2}} \rightarrow \xi_2 = \frac{1}{\xi_1} = \frac{1}{2(z - 2)} \rightarrow w = i\left(\frac{1 + \xi_2}{1 - \xi_2}\right),$$

or

$$w = i\left(\frac{2z - 3}{2z - 5}\right).$$

We know from Theorem 6.3 that this mapping should transform the boundary of D to the boundary of Δ (that is, for each z, $|z - 2| = \frac{1}{2}$, w should be real). To verify this, let $z - 2 = e^{i\theta}/2$, $0 < \theta < 2\pi$, be any point on $\{z:|z - 2| = \frac{1}{2}\}$ except $z = \frac{5}{2}$. A few manipulations reveal that the corresponding $w = (\sin\theta)/(1 - \cos\theta)$ is indeed real; furthermore, when $0 < \theta \leq \pi$, $w \geq 0$, and when $\pi \leq \theta < 2\pi$, $w \leq 0$. If $\theta > 0$ and $\theta \to 0$, $w \to \infty$, and if $\theta < 2\pi$ and $\theta \to 2\pi$, $w \to -\infty$. So $w = i[(2z - 3)/(2z - 5)]$ maps the lower half of $\{z:|z - 2| = \frac{1}{2}\}$ onto the non-positive real axis and the upper half of $\{z:|z - 2| = \frac{1}{2}\}$ onto the non-negative real axis.

Open Wedge onto Upper Half Plane

Let D_α be the domain $\{z:|z| > 0, 0 < \arg z < \alpha\}$ where $\alpha < \pi$. To map D_α 1-1 and conformally onto $\Delta = \{w:\text{Im } w > 0\}$ (see Fig. 6.5), we choose as our mapping function any branch of the function $f(z) = z^{\pi/\alpha}$ with branch cut lying outside D_α. Under the mapping $w = z^{\pi/\alpha}$, points on the boundary of D_α all have real images and those with $\arg z = \alpha$ map onto the negative u-axis, those with $\arg z = 0$ map onto the positive u-axis, and 0 maps onto 0.

Similarly, if $\beta > 0$ and $0 < \alpha < \pi$ are fixed constants, the domain $D_{\alpha,\beta} = \{z:|z| > 0, \beta < \arg z < \alpha + \beta\}$ is mapped 1-1 and conformally onto Δ by any branch of the function

$$f(z) = \begin{cases} z^{\pi/\alpha-\beta} & \text{if } \dfrac{\pi}{\alpha} - \beta > 0 \\ z^{\pi/\alpha+(2\pi-\beta)} & \text{if } \dfrac{\pi}{\alpha} - \beta < 0 \end{cases},$$

whose branch cut lies outside $D_{\alpha,\beta}$.

Example 6.7 The mapping $w = z^2$ transforms $D_{\pi/2} = \{z:|z| > 0, 0 < \arg z < \pi/2\}$ 1-1 and conformally onto $\Delta = \{w:\text{Im } w > 0\}$. For positive numbers c, the level curves $xy = c$ in $D_{\pi/2}$ are mapped onto the straight lines $v = c$ in Δ.

Infinite Strip onto Upper Half Plane

Let D be the infinite strip $\{z:0 < \text{Im } z < \pi\}$ and Δ be the upper half plane $\{w:\text{Im } w > 0\}$. The function $f(z) = e^z$ maps D 1-1 and conformally onto Δ.

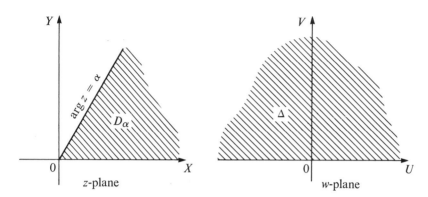

Figure 6.5

If we let $D_1 = \{z:\text{Re } z < 0, 0 < \text{Im } z < \pi\}$, and $D_2 = \{z:\text{Re } z > 0, 0 < \text{Im } z < \pi\}$, then under the mapping $w = e^z$, D_1 is transformed to $\Delta_1 = \{w:0 < |w| < 1, 0 < \arg w < \pi\}$, and D_2 is transformed to $\Delta_2 = \{w:|w| > 1, 0 < \arg w < \pi\}$. Note that the inverse of this mapping is $z = \log w$, where $0 < \arg w < \pi$.

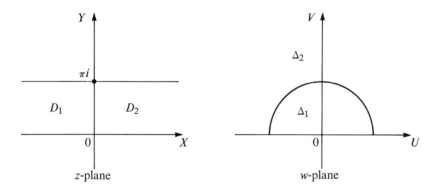

Figure 6.6

Example 6.8 We can alter the mapping above in order to map any infinite strip onto the upper half plane by first transforming the given strip into the one shown in Fig. 6.6.

For example, if $D = \{z:1 < \text{Re } z < 2\}$, then $\xi_1 = iz$ maps D 1-1 and conformally onto $D_1 = \{\xi_1:1 < \text{Im } \xi_1 < 2\}$; $\xi_2 = \pi(\xi_1 - i)$ maps D_1 1-1 and conformally onto $D_2 = \{\xi_2:0 < \text{Im } \xi_2 < \pi\}$; $w = e^{\xi_2}$ maps D_2 1-1 and conformally onto the upper half plane, Δ. Schematically we have

$$(z \in D) \rightarrow (iz \in D_1) \rightarrow [\pi i(z - 1) \in D_2] \rightarrow \{w = \exp[\pi i(z - 1)] \in \Delta\}$$

What is the inverse mapping of Δ onto D in this case?

Semi-infinite Strip onto Upper Half Plane

Let D be the semi-infinite strip $\{z: -\pi/2 < \operatorname{Re} z < \pi/2, \operatorname{Im} z > 0\}$ and Δ be the upper half plane $\{w: \operatorname{Im} w > 0\}$. The mapping $w = \sin z$ maps D 1-1 and conformally onto Δ.

We know that $f(z) = \sin z$ has a nonzero derivative at each point of D; also $w = \sin z = \sin(x + iy) = \sin x \cosh y + i \cos x \sinh y$. For $z \in D$, $\cos x > 0$ and $\sinh y > 0$, so that $w = \sin z \in \Delta$.

To show that $w = \sin z$ is a 1-1 mapping of D onto Δ, let K be any positive number and C_K be the closed path bounding the region $D_K = \{z: -\pi/2 < \operatorname{Re} z < \pi/2, 0 < \operatorname{Im} z < K\}$. It is easy to verify that $w = \sin z$ maps C_K onto the closed path Γ_K in the w-plane shown in Fig. 6.7. Since $w = \sin z$ describes Γ_K once in the positive direction as z goes about C_K once in the positive direction, Theorem 6.2 implies that $w = \sin z$ maps D_K 1-1 onto the region enclosed by Γ_K. If we allow K to increase, we can see that $w = \sin z$ maps D 1-1 onto Δ.

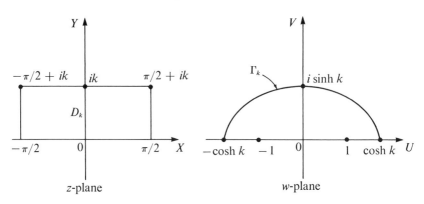

Figure 6.7

Example 6.9 To map $D = \{z: \operatorname{Re} z > 0, 0 < \operatorname{Im} z < 3\}$ 1-1 and conformally onto $\Delta = \{w: \operatorname{Im} w > 0\}$, we can use $\xi_1 = iz$ to map D 1-1 and conformally onto $D_1 = \{\xi_1: -3 < \operatorname{Re} \xi_1 < 0, \operatorname{Im} \xi_1 > 0\}$, $\xi_2 = (\pi/3)\left(\xi_1 + \dfrac{3}{2}\right)$ to map D_1 1-1 and conformally onto $D_2 = \{\xi_2: -\pi/2 < \operatorname{Re} \xi_2 < \pi/2, \operatorname{Im} \xi_2 > 0\}$, and $w = \sin \xi_2$ to map D_2 1-1 and conformally onto Δ. We arrive at the mapping we want by the sequence

$$(z \in D) \rightarrow (iz \in D_1) \rightarrow \left[\left(\frac{\pi}{3}\right)\left(iz + \frac{3}{2}\right) \in D_2 \right]$$

$$\rightarrow \left\{ w = \sin\left[\left(\frac{\pi}{3}\right)\left(iz + \frac{3}{2}\right) \right] \in \Delta \right\}$$

Schwarz-Christoffel Transformations

The Schwarz-Christoffel transformations are mappings which map the upper half plane 1-1 and conformally onto simply connected domains bounded by closed polygonal paths. As such they provide a fairly general answer to the problem of constructing explicitly mappings whose existence is guaranteed by the Riemann mapping theorem. The transformations are given in the form of integrals which, except for simple cases, cannot be evaluated by elementary means. The mathematics of their derivation is by now within the reader's background, but the proof of the statement we make would be more fairly described as a non-elementary use of elementary mathematics. Thus we shall not try to prove what we say; instead we shall suggest by example how the transformations may apply in mapping regions which can be treated as limiting cases of regions bounded by closed polygonal paths.

Let Δ be the set in the w-plane which is the interior of a closed polygonal path with vertices $w_1, w_2, w_3, \ldots, w_n, w_{n+1} = w_1$, as drawn in Fig. 6.8. If $\alpha_1, \alpha_2, \alpha_3, \ldots, \alpha_n$ are the indicated exterior angles at each vertex of the polygon, then $\alpha_1 + \alpha_2 + \cdots + \alpha_n = 2\pi$. We let $x_1 < x_2 < \cdots < x_n$ be the points on the real axis in the z-plane whose images are to be $w_1, w_2, \ldots, w_n$, respectively. Then a Schwarz-Christoffel transformation which maps $D = \{z : \text{Im } z > 0\}$ 1-1 and conformally onto Δ in such a way that x_j is mapped to w_j, $j = 1, 2, \ldots, n$, is given by

$$w = A \int \frac{dz}{(z - x_1)^{\alpha_1/\pi}(z - x_2)^{\alpha_2/\pi} \cdots (z - x_n)^{\alpha_n/\pi}} + B \qquad (6.1)$$

where A and B are constants whose values are determined by the parameters $x_j, w_j, \alpha_j, j = 1, 2, \ldots, n$. If n, the number of vertices, is greater than 4, the choice of $x_1, x_2, \ldots, x_n$ is not completely free; if $n \leq 4$, the points x_1, x_2, x_3, x_4 can be chosen arbitrarily.

Example 6.10   If Δ is the interior of a rectangle with vertices w_1, w_2, w_3, w_4, and if we choose $-1, -\frac{1}{2}, \frac{1}{2}, 1$ as real points to correspond to these vertices, then the angles $\alpha_1, \alpha_2, \alpha_3, \alpha_4$ are all $\pi/2$, and the mapping (6.1) takes the form

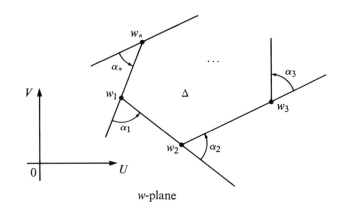

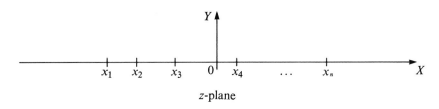

z-plane

Figure 6.8

$$w = A \int \frac{dz}{\sqrt{(z^2 - \tfrac{1}{4})(z^2 - 1)}} + B$$

Even with a geometrically "nice" Δ we obtain a non-elementary mapping function.

With the aid of a preliminary mapping we can effectively reduce by one the number of denominator factors in the integrand of (6.1). The mapping $\xi = -1/(z - x_n)$ maps D 1-1 and conformally onto itself, and substitution in the integrand of (6.1) yields (at least formally)

$$\frac{dz}{(z - x_1)^{\alpha_1/\pi}(z - x_2)^{\alpha_2/\pi}\cdots(z - x_n)^{\alpha_n/\pi}}$$

$$= \frac{\dfrac{1}{\xi^2}\,d\xi}{\left(x_n - \dfrac{1}{\xi} - x_1\right)^{\alpha_1/\pi}\left(x_n - \dfrac{1}{\xi} - x_2\right)^{\alpha_2/\pi}\cdots\left(-\dfrac{1}{\xi}\right)^{\alpha_n/\pi}}$$

$$= \frac{d\xi}{(\xi x_n - 1 - \xi x_1)^{\alpha_1/\pi}(\xi x_n - 1 - \xi x_2)^{\alpha_2/\pi}\cdots(\xi x_n - 1 - \xi x_{n-1})^{\alpha_{n-1}/\pi}}$$

$$= \left[\frac{1}{(x_n - x_1)^{\alpha_1/\pi}(x_n - x_2)^{\alpha_2/\pi}\cdots(x_n - x_{n-1})^{\alpha_{n-1}/\pi}}\right]$$

$$\cdot\left[\frac{d\xi}{\left(\xi - \dfrac{1}{x_n - x_1}\right)^{\alpha_1/\pi}\left(\xi - \dfrac{1}{x_n - x_2}\right)^{\alpha_2/\pi}\cdots\left(\xi - \dfrac{1}{x_n - x_{n-1}}\right)^{\alpha_{n-1}/\pi}}\right]$$

$$= M\left[\frac{d\xi}{(\xi - x_1')^{\alpha_1/\pi}(\xi - x_2')^{\alpha_2/\pi}\cdots(\xi - x_{n-1}')^{\alpha_{n-1}/\pi}}\right]$$

This means that instead of (6.1) we can use

$$w = A + (BM)\int\frac{dz}{(z - x_1')^{\alpha_1/\pi}(z - x_2')^{\alpha_2/\pi}\cdots(z - x_{n-1}')^{\alpha_{n-1}/\pi}} \quad (6.2)$$

Example 6.11   Let $D = \{z : \operatorname{Im} z > 0\}$ and $\Delta = \{w : -\pi/2 < \operatorname{Re} w < \pi/2, \operatorname{Im} w > 0\}$. We saw earlier that $z = \sin w$ must map Δ 1-1 and conformally onto D. We will corroborate this result by constructing a Schwarz-Christoffel mapping of D onto Δ. Now Δ is not a set bounded by a closed polygonal path, but we can regard it as a limiting case as $\operatorname{Im} w_3 \to \infty$ of the region Δw_3 in Fig. 6.9.

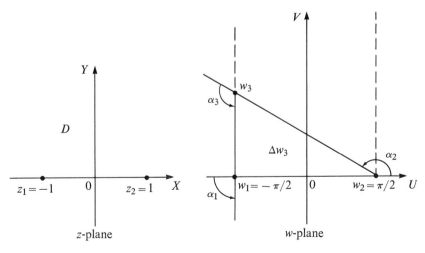

Figure 6.9

Equation (6.2) with $z_1 = -1$, $z_2 = 1$, $\alpha_1 = \pi/2$ gives

$$w = A + (BM)\int \frac{dz}{(z+1)^{1/2}(z-1)^{\alpha_2/\pi}}$$

as a mapping of D onto Δw_3. If Im $w_3 \to \infty$, then $\Delta w_3 \to \Delta$ and $\alpha_2 \to \pi/2$, and we obtain as a limiting case the mapping

$$w = A + (BM)\int \frac{dz}{\sqrt{z^2 - 1}}.$$

That is, for some choice of a branch of $f(z) = \sin^{-1} z$, $w = A + i(BM)\sin^{-1} z$, where A and BM are constants determined by the conditions: $A + i(BM)\sin^{-1}(-1) = -\pi/2$; $A + i(BM)\sin^{-1}(1) = \pi/2$. If we use the principal branch of the inverse sine function, $\sin^{-1}(1) = \pi/2$, $\sin^{-1}(-1) = -\pi/2$, so that $A = 0$, $B = -i$, and $w = \sin^{-1} z$.

For more extensive discussion and examples the reader might see: [Nehari, Chapter 5]; [Pennisi, pp. 393–404]; [Churchill].

PROBLEMS

6.7 Find a 1-1 conformal mapping $w = f(z)$ of D onto Δ in each of the following cases.
  (a) $D = \{z:|z| < 1\}$, $\Delta = \{w:|w| < 1\}$, and $f(\tfrac{1}{2}) = 0$, $f(1) = i$.
  (b) $D = \{z:|z| < 1\}$, $\Delta = \{w:\text{Im } w > 0\}$, and $f(0) = 2i$.
  (c) $D = \{z:\text{Re } z > 0\}$, $\Delta = \{w:\pi < \arg w < 2\pi\}$.
  (d) $D = \{z:|z| > 0, 0 < \arg z < \pi/4\}$, $\Delta = \{w:|w| < 1\}$.
  (e) $D = \{z:-\pi/2 < \text{Re } z < \pi/2, \text{Im } z > 0\}$, $\Delta = \{w:\text{Re } w > 0,$ Im $w > 0\}$.
  (f) $D = \{z:0 < \text{Re } z + \text{Im } z < 1\}$, $\Delta = \{w:\text{Im } w > 0\}$.
  (g) $D = \{z:|z - 1| < 1\}$, $\Delta = \{w:\text{Re } w > 1\}$.

6.8 Show that the transformation $w = [(z + 1)/(z - 1)]^2$ maps $D = \{z:|z| < 1, \text{Im } z > 0\}$ 1-1 and conformally onto $\Delta = \{w:\text{Im } w > 0\}$.

6.9 Show that the transformation $w = z + 1/z$ maps $D = \{z:|z| > 1,$ Im $z > 0\}$ 1-1 and conformally onto $\Delta = \{w:\text{Im } w > 0\}$.

6.10 Let Δ be the infinite strip $\{w:0 < \text{Im } w < \pi\}$ with a semi-infinite slit along $\{w:\text{Re } w \leq 0, \text{Im } w = \pi/2\}$, as drawn in Fig. 6.10.

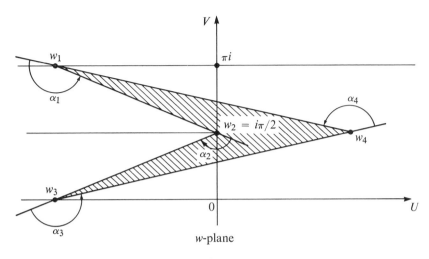

Figure 6.10

If λ is any positive real number, let $w_1 = -\lambda + i$, $w_2 = \pi i/2$, $w_3 = -\lambda$, $w_4 = \lambda + \pi i/2$, and Δ_λ be the shaded open set bounded by the quadrilateral with vertices w_1, w_2, w_3, w_4. Write a Schwarz-Christoffel transformation $w = f(z,\lambda)$ mapping $D = \{z: \text{Im } z > 0\}$ onto Δ_λ with $w_1 = f(-1,\lambda)$, $w_2 = f(0,\lambda)$, $w_3 = f(1,\lambda)$.

As $\lambda \to \infty$, then $\alpha_1 \to \pi$, $\alpha_2 \to -\pi$, $\alpha_3 \to \pi$, and the function $f(z,\lambda)$ approaches $f(z)$, a function which maps D onto Δ 1-1 and conformally. Determine $f(z)$.

6.11 Let D be the domain in the z-plane of points inside the circle $C_1 = \{z: |z| = 1\}$ and outside the circle $C_2 = \{z: |z - \sqrt{3}/4| = \sqrt{3}/4\}$.

   (a) Show that the mapping $w = (z - \sqrt{3})/(\sqrt{3}z - 1)$ maps D 1-1 and conformally onto $\Delta = \{w: 1 < |w| < \sqrt{3}\}$.

   (b) Show that $w = (z - \sqrt{3})/(\sqrt{3}z - 1)$ maps C_1 onto $\{w: |w| = 1\}$ and C_2 onto $\{w: |w| = \sqrt{3}\}$.

6.12 Let $D = \{z: |z| < 1, \text{Im } z > 0\}$. Show that $w = [(1 + z)/(1 - z)]^2$ maps D 1-1 and conformally onto $\Delta = \{w: \text{Im } w > 0\}$.

Section 6.3   Applications to a Boundary-Value Problem

In more than a few branches of the physical sciences an answer is sought to the following problem: if G is a domain in R^n, Γ is its boundary, and

$h(\mathbf{Q})$ is a real-valued function defined for vectors $\mathbf{Q} \in \Gamma$, can one find a function $g(\mathbf{P})$, defined for vectors $\mathbf{P}$ in G for which

(1) $\nabla^2 g = 0$ everywhere in G;
(2) $\lim_{P \to Q} g(\mathbf{P}) = h(\mathbf{Q})$ for each $\mathbf{Q} \in \Gamma$?

In the two-dimensional case the problem is usually called the *Dirichlet problem*, and solutions for it are known if G, Γ, and h are assumed to have certain properties. The solutions in two dimensions rest on the close relation of harmonic functions to analytic functions, and the properties of conformal mappings. Thus the methods of solution in the two-dimensional case are not generally useful in higher-dimensional spaces.

We shall first treat the Dirichlet problem for $G = \{z:\text{Im } z > 0\}$, stating the main result without proof. Then we shall examine the problem for other simply connected domains. For a far more extensive treatment of this and related two-dimensional problems, [Carrier, Chapter 4] is a very good reference.

The Upper Half Plane

If $G = \{z:\text{Im } z > 0\}$ and $\Gamma = \{z:\text{Im } z = 0\}$, and $h(z)$ is a real-valued, bounded, piecewise continuous function defined on Γ (that is, $h(z)$ is continuous at all but finitely many points of Γ) the Dirichlet problem for G, Γ, and h has a unique bounded solution. This solution is represented with an improper integral by what is known as the *Poisson integral formula for the upper half plane*.

This is the theorem we shall use without proof.

THEOREM 6.5 Hypotheses: $G = \{z:\text{Im } z > 0\}$, $\Gamma = z:\text{Im } z = 0\}$, and $h(z)$ is real-valued, bounded, and piecewise continuous on Γ.

Conclusions:
C1 For each $z = x + iy$ in G, the improper integral

$$g(x,y) = \frac{1}{\pi} \int_{-\infty}^{\infty} \frac{y\, h(r)}{(x - r)^2 + y^2}\, dr$$

converges.
C2 $g(x,y)$ is harmonic in G.
C3 For any real number M such that $|h(z)| \le M$ on Γ, $|g(x,y| \le M$ everywhere in G.
C4 For each real number r at which $h(z)$ is continuous,

$$\lim_{(x,y)\to(r,0)} g(x,y) = h(r).$$

C5 $g(x,y)$ is the unique bounded function in G satisfying C2–C4.

It is not feasible, in general, to determine $g(x,y)$ explicitly from this integral formula. However, some techniques of numerical integration might be used to approximate values of $g(x,y)$, and for some choices of $h(z)$ we can determine $g(x,y)$ completely. Theorem 6.5 and the Poisson integral formula can also be stated for the open unit disk, and these will appear in Chapter 7.

In the case where $h(z)$ is a "step" function — that is, $h(z)$ is constant on open intervals of the real axis except for a finite number of "jumps" — we can determine explicit solutions to the Dirichlet problem. At the jump discontinuities of $h(z)$, C4 of Theorem 6.5 will not apply. The boundary behavior of the solution at such points is more complicated, and problem 6.14 partially explains what happens.

Example 6.12   Let

$$h(z) = \begin{cases} 1, & z < 0 \\ 0, & z > 0 \end{cases} \quad \text{on} \quad \Gamma.$$

Then for $z = x + iy$ in G,

$$g(x,y) = \frac{1}{\pi} \int_{-\infty}^{0} \frac{y}{(x-r)^2 + y^2}\, dr = \frac{1}{\pi}\left(\frac{\pi}{2} - \tan^{-1}\left(\frac{x}{y}\right)\right)$$

$$= \frac{1}{\pi} \tan^{-1}\left(\frac{y}{x}\right) = \frac{\text{Arg } z}{\pi},$$

where we have used the principal branch of $\tan^{-1} z$. It is easy to verify that $g(x,y)$ is harmonic in G and converges to $h(r)$ as $(x,y) \to (r,0)$ for every nonzero r.

Now let us take a more general step-function with $n + 1$ different values, for some positive integer n.

Let $r_1 < r_2 < r_3 < \cdots < r_n$ and $k_0, k_1, k_2, \ldots, k_n$ be arbitrary real constants, and define

$$h(z) = \begin{cases} k_0, & z < r_1 \\ k_j, & r_j < z < r_{j+1}, \quad j = 1, 2, \ldots, n-1 \\ k_n, & r_n < z \end{cases}$$

Then

$$g(x,y) = \frac{1}{\pi}\left\{ \int\limits_{-\infty}^{r_1} \frac{k_0 y}{(x-r)^2 + y^2}\, dr + \sum_{j=1}^{n-1} \int\limits_{r_j}^{r_{j+1}} \frac{k_j y}{(x-r)^2 + y^2}\, dr \right.$$

$$\left. + \int\limits_{r_n}^{\infty} \frac{k_n y}{(x-r)^2 + y^2}\, dr \right\}.$$

As the reader can verify,

$$\int\limits_{r_n}^{\infty} \frac{k_n y}{(x-r)^2 + y^2}\, dr = k_n\, [\pi - \text{Arg}(z - r_n)],$$

and for each $j = 1, 2, \ldots, n-1$,

$$\int\limits_{r_j}^{r_{j+1}} \frac{k_j y}{(x-r)^2 + y^2}\, dr = k_j \left[\text{Arg}\!\left(\frac{z - r_{j+1}}{z - r_j} \right) \right].$$

Also,

$$\int\limits_{-\infty}^{r_1} \frac{k_0 y}{(x-r)^2 + y^2}\, dr = k_0\, \text{Arg}(z - r_1).$$

Consequently, for $z = x + iy$ in G,

$$\begin{aligned} g(x,y) &= \frac{1}{\pi}\left\{ k_0\, \text{Arg}(z - r_1) + \sum_{j=1}^{n-1} k_j\, \text{Arg}\!\left(\frac{z - r_{j+1}}{z - r_j} \right) + k_n[\pi - \text{Arg}(z - r_n)] \right\}. \end{aligned}$$

Example 6.13   In particular, if

$$h(z) = \begin{cases} 0, & z < -1 \\ \pi, & -1 < z < 1 \\ 0, & 1 < z \end{cases},$$

we have $g(z) = g(x,y) = \text{Arg}[(z - 1)/(z + 1)]$.

Other Simply Connected Domains

Suppose D is a simply connected domain with boundary C and satisfies the hypotheses of the Riemann mapping theorem. Suppose we let $f(z)$ be a function analytic in D which maps D 1-1 and conformally onto $G = \{w:\text{Im } w > 0\}$ and maps C 1-1 onto $\Gamma = \{w:\text{Im } w = 0\}$. To solve a Dirichlet problem for D and C, we use the mapping $w = f(z)$ to consider an associated Dirichlet problem for G and Γ. If we can solve this associated Dirichlet problem, we can use the'inverse of the mapping $w = f(z)$ to determine a solution to the original Dirichlet problem for D and C.

The validity of this method rests on two theorems whose proofs depend on elementary properties of analytic functions and mappings by analytic functions.

THEOREM 6.6   Hypotheses:
H1 D is a simply-connected domain in the z-plane.
H2 $w = f(z)$ is an analytic function mapping D into an open set G of the w-plane.
H3 $g(w)$ is a real-valued function harmonic in G.

Conclusion: $G(z) = g[f(z)]$ is harmonic in D.

Proof: Let z_0 be any point of D and let $w_0 = f(z_0)$ in G. Choose $\rho > 0$ so small that $\{w:|w - w_0| < \rho\}$ is contained in G. Since $f(z)$ is continuous at z_0 we can find a number $r > 0$ such that $N(z_0,r) = \{z:|z - z_0| < r\}$ is contained in D and $w = f(z)$ maps $N(z_0,r)$ into $\{w:|w - w_0| < \rho\}$.

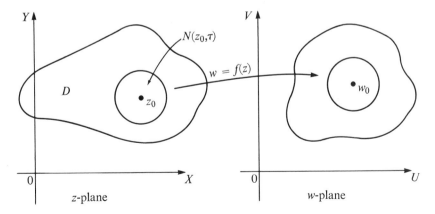

Figure 6.11

By Theorem 1.8 we can find an analytic function $F(w)$, with $g(w) =$ Re $F(w)$ for $\{w:|w - w_0| < \rho\}$. Then $F[f(z)]$ is analytic in $N(z_0,r)$, and $g[f(z)] =$ Re $F[f(z)]$ is harmonic in $N(z_0,r)$. Since z_0 is an arbitrary point of D, $g[f(z)] = G(z)$ is harmonic in D.

If the mapping of this theorem is also 1-1 and conformal in D, we have a stronger conclusion.

THEOREM 6.7   Hypotheses:

H1 D is a simply-connected domain with path C as its boundary, and $f(z)$ is an analytic function which maps D 1-1 and conformally onto $G = \{w:\text{Im } w > 0\}$.

H2 $g(w)$ is a real-valued function harmonic in G, and $h(w)$ is a real-valued function defined on $\Gamma = \{w:\text{Im } w = 0\}$.

H3   At each point $w_0 \in \Gamma$ where $h(w)$ is continuous, $\lim_{w \to w_0(w \in G)} g(w) = h(w_0)$.

Conclusions:

C1 $G(z) = g[f(z)]$ is harmonic in D, and $H(z) = h[f(z)]$ is defined on C.

C2   At each point $z_0 \in C$ where $H(z)$ is continuous, $\lim_{z \to z_0(z \in D)} G(z) = H(z_0)$.

Outline of Proof: Theorem 6.6 showed that $G(z)$ is harmonic in D. With an extended version of Theorem 6.3 it can be established that $H(z) = h[f(z)]$ is defined on C and continuous at any point z_0 on C for which $h(w)$ is continuous at $f(z_0)$ on Γ.

If $H(z)$ is continuous at $z = z_0 \in C$ and $w_0 = f(z_0)$, for $z \in D$, $|G(z) - H(z_0)| = |g[f(z)] - h[f(z_0)]| = |g(w) - h(w_0)|$, where $w \in G$. Given any number $\epsilon > 0$, H3 implies that there exists a number $\delta > 0$ such that $|g(w) - h(w_0)| < \epsilon$ whenever $0 < |w - w_0| < \delta$ for $w \in G$. But $|w - w_0| = |f(z) - f(z_0)|$, and we can find a number $\lambda > 0$ such that $|f(z) - f(z_0)| < \delta$ when $0 < |z - z_0| < \lambda$ for $z \in D$. (Why is it true that $0 < |f(z) - f(z_0)|$ for $z \in D$, $z_0 \in C$?) Then for $z \in D$, $0 < |z - z_0| < \lambda$, we have $|G(z) - H(z_0)| < \epsilon$.

With Theorem 6.7 we see that, by mapping D 1-1 and conformally onto G, solutions to Dirichlet problems for G and Γ determine for us solutions to Dirichlet problems for D and C as long as the boundary values we consider on C are continuous for all but some isolated points on C.

If D is a simply-connected domain with a path C as its boundary and satisfies the hypotheses of the Riemann mapping theorem, and if $h(z)$ is a real-valued function defined and bounded on C, then we know we can find a solution to the Dirichlet problem for D, C, and h if h is piecewise continuous on C. If h is continuous on C this solution will be unique; if h is piecewise continuous on C, there will not be a unique solution, but there will be a unique bounded solution. For less well-behaved boundary functions, discussion of solutions to Dirichlet problems requires more sophisticated analysis of real-valued functions and their integrals.

Example 6.14   Let $D = \{z : \operatorname{Re} z > 0, \operatorname{Im} z > 0\}$ and C be its boundary. Suppose

$$H(z) = \begin{cases} 0, & \text{if } z \in C, \quad \operatorname{Re} z = 0 \\ 1, & \text{if } z \in C, \quad \operatorname{Im} z = 0 \end{cases}.$$

The mapping $w = z^2$ transforms D to $G = \{w : \operatorname{Im} w > 0\}$, with boundary $\Gamma = \{w : \operatorname{Im} w = 0\}$. We solve the Dirichlet problem for G, Γ, and the boundary function $h(w) = \begin{cases} 0, & w < 0 \\ 1, & w > 0 \end{cases}$; we obtain $g(w) = (1/\pi)(1 - \arg w)$. Using Theorem 6.7 we know $G(z) = g(z^2) = (1/\pi)[1 - \arg(z^2)]$ is harmonic in D and has boundary values $H(z)$ on C.

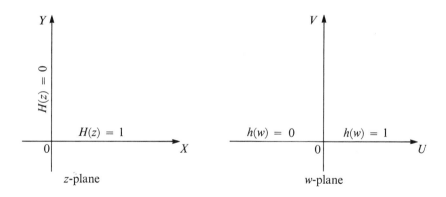

Figure 6.12

Example 6.15   Let D be the z-plane with a slit along the non-negative real axis, so that $D = \{z : |z| > 0, 0 < \arg z < 2\pi\}$.
We define $H(z)$ on the non-negative real axis by

$$H(z) = \begin{cases} 2, & \arg z = 0 \\ -2, & \arg z = 2\pi \end{cases}.$$

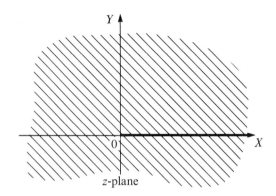

Figure 6.13

We seek a function $G(z)$, harmonic in D, such that for any real number $s > 0$, $\lim_{z \to s(z \in D)} G(z) = \pm 2$, where "$+$" or "$-$" is chosen depending upon whether z approaches s from above or below the positive real axis.

Now the branch of $w = z^{1/2}$ defined by $w = r^{1/2} \exp(i\theta/2)$, $r > 0$, $0 < \theta < 2\pi$, maps D 1-1 and conformally onto the upper half plane G with boundary the real axis Γ. The corresponding boundary function is $h(w) = \begin{cases} -2, & w < 0 \\ 2, & w > 0 \end{cases}$, and the corresponding harmonic function in G is $g(w) = (1/\pi)(2\pi - 4 \operatorname{Arg} w)$. So we want $G(z) = (1/\pi)(2\pi - 4 \operatorname{Arg} z^{1/2}) = (2/\pi)(\pi - \operatorname{Arg} z)$.

PROBLEMS

6.13 Solve the Dirichlet problems for each of the following regions with boundary functions indicated.

(a) $D = \{z : \operatorname{Re} z > 0, \operatorname{Im} z > 0\}$,

$$h(z) = \begin{cases} 0, & z = iy, \quad y > 1 \\ 1, & z = iy, \quad 0 < y < 1 \\ 1, & z = x, \quad 0 < x < 1 \\ 0, & z = x, \quad x > 1 \end{cases}$$

(b) $D = \{z : |z| > 0, 0 < \arg z < \pi/4\}$,

$$h(z) = \begin{cases} 1, & z = r \exp(i\pi/4), \quad r > 0 \\ 0, & z = r, \quad r > 0 \end{cases}$$

(c) $D = \{z:0 < \text{Im } z < 1\}$,

$$h(z) = \begin{cases} 0, & \text{Re } z < 0, \quad \text{Im } z = 0, 1 \\ 3, & \text{Re } z > 0, \quad \text{Im } z = 0, 1 \end{cases}$$

(d) $D = \{z:-\pi/2 < \text{Re } z < \pi/2, \text{Im } z > 0\}$,

$$h(z) = \begin{cases} 1, & z = \pm(\pi/2) + iy, \quad y > 0 \\ 0, & z = x, \quad -\pi/2 < x < \pi/2 \end{cases}$$

(e) $D = \{z:|z| > 1, 0 < \arg z < \pi\}$,

$$h(z) = \begin{cases} 1, & z = x, \quad |x| > 1 \\ -1, & z = e^{i\theta}, \quad 0 < \theta < \pi \end{cases}$$

(f) $D = \{z:|z| < 1\}$, $h(z) = \begin{cases} 0, & |z| = 1, \quad -\pi < \arg z < 0 \\ 1, & |z| = 1, \quad 0 < \arg z < \pi \end{cases}$

6.14 Suppose $g(z)$ is a solution to the Dirichlet problem for the upper half plane with boundary function $h(z) = \begin{cases} k_1, & z < r_1 \\ k_2, & z > r_1 \end{cases}$. Show that for $y > 0$, $\lim_{y \to 0} g(r_1 + iy) = (k_1 + k_2)/2$. Also evaluate for $s > 0 \lim_{s \to 0} g(r_1 + s \exp(i\pi/4))$.

6.15 Use the techniques of Chapter 5 to determine

$$g(x,y) = \frac{1}{\pi} \int_{-\infty}^{\infty} \frac{y \, h(r)}{(x - r)^2 + y^2} \, dr$$

when $h(r) = \cos r$.

6.16 Let D be the infinite strip $\{z:0 < \text{Im } z < \pi\}$ with a slit along the line $L_3 = \{z:z = x + i\pi/2, x \leq 0\}$ as drawn in Fig. 6.10. The boundary C of D consists of the lines $L_1 = \{z:\text{Im } z = \pi\}$, $L_2 = \{z:\text{Im } z = 0\}$, and L_3. Find a real-valued function $G(z)$, harmonic in D, such that as $z \in D$ approaches any point on

$$\begin{bmatrix} L_1 \\ L_2 \\ L_3 \end{bmatrix}, \quad G(z) \text{ has limit } \begin{bmatrix} 1 \\ -1 \\ 0 \end{bmatrix}.$$

(From problem 6.10, a branch of $w = (1 + e^{2z})^{1/2}$ maps D 1-1 and conformally onto $G = \{w:\text{Im } w > 0\}$.)

6.17 Let $D = \{z:|z| < 1, \text{Im } z > 0\}$. Define $h(z)$ on the boundary of D by

$$h(z) = \begin{cases} 0, & |z| = 1, \ \frac{\pi}{2} < \arg z < \pi \\ 1, & |z| = 1, \ 0 < \arg z < \frac{\pi}{2} \\ 0, & -1 < z < 0 \\ 1, & 0 < z < 1 \end{cases}$$

Give a solution to the Dirichlet problem for D and its boundary with boundary values $h(z)$.

6.18 Suppose a and b are real numbers, $0 < a < b$, and $D = \{z : a < |z| < b\}$, $C_b = \{z : |z| = b\}$, $C_a = \{z : |z| = a\}$. Define $h(z)$ on the boundary of D by $h(z) = \begin{cases} B, & \text{if } z \in C_b \\ A, & \text{if } z \in C_a \end{cases}$, where A and B

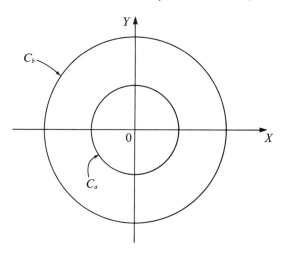

Figure 6.14

are constants. Verify that

$$g(z) = A + (B - A)[(\text{Log } |z| - \text{Log } a)/(\text{Log } b - \text{Log } a)]$$

gives a solution of the Dirichlet problem for D and its boundary with boundary values $h(z)$.

6.19 Let D, C_1 and C_2 be as in problem 6.11. Use the mapping of that problem and the result of problem 6.18 to find a function $g(z)$, harmonic in D, such that

$$\lim_{z \to \xi (z \in D)} g(z) = \begin{Bmatrix} A, & \xi \in C_1 \\ B, & \xi \in C_2 \end{Bmatrix},$$

where A and B are arbitrary constants.

Section 6.4   Green's Functions and a Generalized Dirichlet Problem

Let $D = \{z: |z| < 1\}$ and $C = \{z: |z| = 1\}$. If $h(z)$ is a real-valued function defined on C, and z_0 is a fixed point in D, and ρ is a fixed real number, we can pose a more general Dirichlet problem. Find a real-valued function $g(z)$ such that:

   (1) $g(z)$ is defined and harmonic at all points of D except $z = z_0$;

   (2) $\lim_{z \to \xi (z \in D)} g(z) = h(\xi)$ at each point $\xi \in C$ where $h(\xi)$ is continuous;

   (3) $\lim_{z \to z_0} g(z)/\log|z - z_0| = \rho$.

It is the condition (3) which makes this problem different from the problems of the previous section. We do not require $g(z)$ to be harmonic at z_0; instead we allow $g(z)$ to become unbounded in the neighborhood of z_0 in a specified fashion.

We solve this problem by considering two separate problems as follows. First suppose we can solve an ordinary Dirichlet problem for D and C with boundary values $h(z)$ (that is, we can find $g_0(z)$, harmonic in all of D, such that $\lim_{z \to \xi (z \in D)} g_0(z) = h(\xi)$ at each point $\xi \in C$ where $h(\xi)$ is continuous). Next we try to find a real-valued function $g(z, z_0)$ such that:

   (1') $g(z, z_0)$ is defined and harmonic in D except for $z = z_0$;

   (2') $\lim_{z \to \xi (z \in D)} g(z, z_0) = 0$ for every $\xi \in C$;

   (3') $\lim_{z \to z_0} g(z, z_0)/\log|z - z_0| = -1$.

If we can find these two functions, $g_0(z)$ and $g(z, z_0)$, the function $g(z) = [g_0(z) - \rho g(z, z_0)]$ will have the properties (1), (2), (3) and will solve our problem. (The reader should verify this.)

The function $g(z, z_0)$ is known as a *Green's function for region D with singularity at z_0*. It may appear here as a solution to a very special and technical Dirichlet problem, but its special importance in complex analysis and in solving partial differential equations is more extensive than our uses of it will suggest. In the remainder of this chapter we shall see a connection between the existence of Green's functions for simply connected domains and the existence of a 1-1 conformal mapping of such a domain onto the open unit disk. In Chapter 7 we shall describe one way in which Green's functions are significant in arriving at the solution for a class of partial differential equations boundary value problems.

It is easy to verify that a Green's function for D with singularity at $z = 0$ is given by $g(z,0) = -\log|z|$. For any point z^* in D, the function $f(z) = (z - z^*)/(1 - \overline{z^*}z)$ maps D 1-1 and conformally onto itself with $f(z^*) = 0$. Then a Green's function for D with singularity at z^* is determined by

$$g(z,z^*) = g[f(z),0] = g\left[\left(\frac{z - z^*}{1 - \overline{z^*}z}\right),0\right]$$

$$= -\log\left|\frac{z - z^*}{1 - \overline{z^*}z}\right|.$$

Consequently a solution to the more general Dirichlet problem satisfying (1), (2), (3) is

$$g(z) = g_0(z) + \rho \log\left|\frac{z - z_0}{1 - \overline{z_0}z}\right|$$

where $g_0(z)$ is the solution to an ordinary Dirichlet problem for D, C, and $h(z)$.

Example 6.16   Suppose $h(z) = \begin{cases} 0, & |z| = 1, & -\pi < \arg z < 0 \\ 1, & |z| = 1, & 0 < \arg z < \pi \end{cases}$ and $z_0 = i/2$, $\rho = 2$. From problem 6.13(f) we may take $g_0(z) = (1/\pi) \arg[i(1 + z)/(1 - z)]$, and

$$g(z) = \frac{1}{\pi} \arg\left[\frac{i(1 + z)}{1 - z}\right] + 2 \log\left(\frac{2z - i}{iz + 2}\right).$$

To solve a generalized Dirichlet problem for other simply connected domains is not difficult as long as we can find analytic functions to map these domains 1-1 and conformally onto the open unit disk.

Definition   Let D be any simply-connected domain which is not the entire finite complex plane, C be its boundary, and z_0 be any point of D. A *Green's function for domain D with singularity at z_0* is a real-valued function $g(z,z_0)$, defined in D except at z_0, and having these properties:

(1') $g(z,z_0)$ is harmonic in D except for $z = z_0$;
(2') for every $\xi \in C$, $\lim_{z \to \xi(z \in D)} g(z,z_0) = 0$;
(3') $\lim_{z \to z_0} g(z,z_0)/\log|z - z_0| = -1$.

An obvious question to ask here is whether there is a Green's function $g(z,z_0)$ for a given domain D and a point $z_0 \in D$. In terms of the results we have already, we can give a partial answer.

THEOREM 6.8 Hypotheses:

H1 D is a simply connected domain which is not the entire finite complex plane whose boundary C is a closed path.

H2 z_0 is a fixed point of D, and $f(z)$ is a function analytic in D which maps D 1-1 and conformally onto $\{w:|w| < 1\}$ with $f(z_0) = 0$.

Conclusion: $g(z,z_0) = -\log |f(z)|$ is a Green's function for D with singularity at z_0.

Proof: Since $f(z)$ is analytic and 1-1 in D, we can write $f(z) = (z - z_0)F(z)$, where $F(z)$ is analytic and never equal to 0 in D. In a neighborhood of each point of D we can choose a branch of $\log F(z)$ analytic in that neighborhood; and since $\log |F(z)|$ is the same for all the branches of $\log F(z)$, we see that $\log |F(z)|$ is harmonic in D. Consequently, $g(z,z_0) = -\log |f(z)| = -\log |z - z_0| - \log |F(z)|$ is harmonic in D except for $z = z_0$.

An application of Theorem 6.3 will show that for any point $\xi \in C$, $\lim_{z \to \xi (z \in D)} g(z,z_0) = 0$. And since $F(z_0) \neq 0$, we can see that

$$\lim_{z \to z_0 (z \in D)} \frac{g(z,z_0)}{\log|z - z_0|} = -1.$$

In order to say more about Green's functions we state two theorems without proof. If there is a Green's function for D with singularity at $z_0 \in D$, Theorem 6.9 tells us that it is unique; this suggests that, up to a selection of the point $z_0 \in D$, a Green's function $g(z,z_0)$ for region D somehow represents the region D. In Theorem 6.10 it appears that $g(z,z_0)$ can as well represent all the regions onto which D can be mapped 1-1 and conformally.

THEOREM 6.9 Hypotheses:

H1 D is a simply connected domain which is not the entire finite complex plane, C is its boundary, and z_0 is any point of D.

H2   Both $g(z,z_0)$ and $h(z,z_0)$ are real-valued and satisfy (1'), (2'), (3').

Conclusion: For every point z of D, $z \neq z_0$, $g(z,z_0) = h(z,z_0)$.

THEOREM 6.10 Hypotheses:

H1 $g(z,z_0)$ is a Green's function for a domain D with singularity at $z_0 \in D$.

H2 $f(w)$ is analytic in a domain Δ and maps Δ 1-1 and conformally onto D with $f(w_0) = z_0$.

Conclusion: The function $G(w,w_0) = g[f(w),f(w_0)] = g(z,z_0)$ is a Green's function for Δ with singularity at $w_0 \in \Delta$.

PROBLEMS

6.20 Find a Green's function for the region $\{z : \text{Im } z > 0\}$ with singularity at $z = i\alpha$, $\alpha > 0$.

6.21 Find a function $g(z)$, harmonic in the upper half plane except for $z = i$, such that:

(a) for $y > 0$, $\lim_{y \to 0} g(k + iy) = \begin{cases} 2, & \text{if } k < 0 \\ -2, & \text{if } k > 0 \end{cases}$; and

(b) $\lim_{z \to i} [g(z)/\log |z - i|] = 2$.

6.22 If $g(z,i)$ is a Green's function for the upper half plane with singularity at $z = i$, verify that the level curves $g(z,i) = k$, for $k > 0$, are circles lying in the upper half plane with centers at $z = 0 + i \coth k$ and radii csch k. What happens to the level curve $g(z,i) = k$ as $k \to \infty$?

6.23 Find a Green's function for the region $D = \{z : |z| > 1, \text{Im } z > 0\}$ with singularity at $z = 2i$.

6.24 Find a function $g(z)$, harmonic for $0 < |z| < 1$, such that:

(a) $\lim_{z \to e^{i\theta}} g(z) = \begin{cases} -1, & \text{if } -\pi < \theta < 0 \\ 1, & \text{if } 0 < \theta < \pi \end{cases}$; and

(b) $\lim_{z \to 0} [g(z)/\log |z|] = -3$.

(See problem 6.13(f).)

6.25 Find a function $g(z)$, harmonic in $D = \{z : |z - i| < 2\}$ except for $z = i$ and $z = 0$, which has the following properties:

(a) $\lim_{z \to \xi} g(z) = \begin{cases} 2, & \text{if } -\pi < \arg[(\xi - i)/2] < 0 \\ -2, & \text{if } 0 < \arg[(\xi - i)/2] < \pi \end{cases}$;

(b) $\lim_{z \to 0} [g(z)/\log |z|] = -1$;

(c) $\lim_{z \to i} [g(z)/\log |z - i|] = 1$.

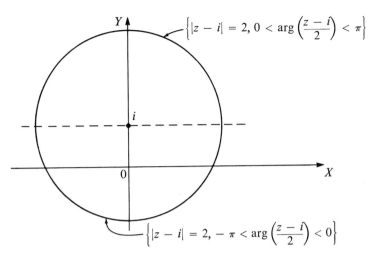

$$\left\{|z - i| = 2,\, 0 < \arg\left(\frac{z - i}{2}\right) < \pi\right\}$$

$$\left\{|z - i| = 2,\, -\pi < \arg\left(\frac{z - i}{2}\right) < 0\right\}$$

Figure 6.15

6.26 If $D = \{z : |z| < 1, \text{Im } z > 0\}$, find a function $g(z)$, harmonic in D except for $z = i/2$, such that

(a) $\lim_{z \to \xi (z \in D)} g(z) = \begin{cases} 1, & |\xi| = 1, \quad 0 < \arg \xi < \pi \\ -1, & -1 < \xi < 1 \end{cases}$;

(b) $\lim_{z \to i/2} [g(z)/\log |z - i/2|] = -\frac{1}{2}$.

(Use problem 6.12 and Theorem 6.10 to find a Green's function for D with singularity at $z = i/2$. Also see problem 6.17.)

7

Two Applications
of Conformal Mapping

Conformal mapping techniques can be of help in a variety of two
dimensional physical problems — or, more precisely, mathematical
models for physical problems. The chief help they offer to such problems
is to change the shape of the region in which the problem is posed to one
in which the problem might be more easily solved. In this chapter we
mention two kinds of problems where conformal mappings can help.
In the first kind of problem the conformal mappings alone can reduce
the problem to a very simple problem, solvable by inspection. In the
second kind of problem, the influence of conformal mappings is in
finding Green's functions for plane domains, and the actual problem-
solving still involves considerable effort.

Section 7.1   Complex Potentials

Let D be a simply connected domain in the two dimensional plane, and
let F be a continuous vector field on D (for each $z \in D$, $F(z)$ is a two
dimensional vector, and $F(z)$ is continuous in D). The field is equivalently
described by defining two continuous, real-valued functions in D to serve
as the components of F. If these component functions have continuous
first partial derivatives, then the field F is called *continuously differentiable*.

Definition   A force field F, continuously differentiable in D, is called
conservative in D if the following conditions hold throughout D:
   (1) $\nabla \times F = 0$;
   (2) $F = \nabla f$ for a real-valued function $f(z)$ with continuous second
         partial derivatives throughout D. (It is proved in advanced
         calculus that (1) and (2) are equivalent.)

Definition   A force field F, continuously differentiable in D, is called
solenoidal in D if $\nabla \cdot F = 0$ throughout D.

If F is conservative in D, then any function f such that $\nabla f = F$ in D is
called a *potential* of F. The fields in which we are directly interested now
are the vector fields over D which are both conservative and solenoidal.
If f is a potential for such a field, F, then $F = \nabla f$ and $\nabla \cdot F = \nabla \cdot \nabla f = \nabla^2 f = 0$ in D. Potentials for such fields are harmonic functions.

In the present terminology, the Dirichlet problems in Chapter 6 were
all asking for potentials in a region having prescribed boundary values.
In this section we try to describe the vector field instead of its potentials.
As long as D is simply connected, for each potential f of F there is a
potential g conjugate to f such that $h(z) = f(z) + i\,g(z)$ is analytic in D.
We call $h(z)$ a *complex potential of F*.

Note that if f is a potential of F and $h(z)$ is a complex potential of F,
then for any real constant k, $f + k$ is also a potential of F; and for any
complex constant w, $h(z) + w$ is also a complex potential of F.

THEOREM 7.1   Hypotheses:
H1 D is a simply connected domain.
H2 F is a conservative solenoidal force field in D.

Conclusion: If $h(z)$ is any complex potential of F in D, then for each
$z \in D$,

$$F(z) = \overline{h'(z)}.$$

Proof: If $h(z) = f(z) + i\ g(z)$ is a complex potential for **F** (which means $h(z)$ is analytic in D and $\nabla(\text{Re } h) = \nabla f = \textbf{F}$ in D) then $h'(z) = f_x + i\ g_x$. From the Cauchy-Riemann equations, we know that $g_x = -f_y$ at each point of D, so that $h'(z) = f_x - if_y$, and $\overline{h'(z)} = f_x + if_y$. But at each point of D, $\textbf{F}(z) = \nabla f = (f_x, f_y)$, so the vectors **F** and $\overline{h'(z)}$ are identical in D.

One way to represent a vector force field **F** is to plot at each point the vector value of **F** at that point. But these plots are tedious to make and possibly hard to interpret. Another way to describe the field **F** is to find the level curves of a potential for **F** — to describe the curves $\{(x,y) : f(x,y) = k\}$ where $\textbf{F} = \nabla f$ and k assumes various values. If $h(z)$ is a complex potential for **F**, then Re $h(z)$ is a potential of **F**, and a more complete description of **F** results from describing the level curves of both Re $h(z)$ and Im $h(z)$. These last functions, being harmonic conjugates, have pairwise orthogonal families of level curves (problem 1.31.)

These families of level curves are given various names depending on the area of application in which they are used. We borrow the language of fluid-flow problems to make this definition.

Definition Let **F** be a conservative solenoidal vector field in a simply connected (plane) domain D. Let $h(z) = f + ig$ be a complex potential for **F** in D. The level curves of f are called *equipotentials* of **F**, and the level curves of g are called *streamlines* of **F**.

Since $\textbf{F} = \nabla f$ in D, the vector field at each point of D is perpendicular to the equipotential through that point and is directed along the streamline through that point.

Let us place our discussion in a simplified physical setting by talking of fluid flows and their velocity fields. Assume that an ideal (incompressible, nonviscous) fluid is flowing through a three dimensional region with the XY-plane as a plane of symmetry so that:

(1) the flow is exactly the same in any plane parallel to the XY-plane.

(2) the velocity vector **F** of the flow at each point is the constant vector $(k,0)$, where k is a positive real number.

What we have is the simplest example of a uniform, two dimensional fluid flow (to the right). A complex potential for the velocity field $\textbf{F} = (k,0)$ of this flow is $h(z) = kz$. The equipotentials are vertical lines, and the streamlines are horizontal lines in the plane.

Now, in theory, any simply-connected domain can be mapped 1-1 and conformally onto the upper half plane. Such a mapping and its inverse preserve the property of being a harmonic function, which is the key functional property involved in potential fields. Then, in principle at least, we should be able to find the velocity field for any uniform, two dimensional flow through any region whose projection on the XY-plane is a domain satisfying the hypotheses of the Riemann mapping theorem. The method is this: Map the domain onto the upper half plane where a uniform flow has a complex potential $h(z) = kz$; then make a change of variables to determine a complex potential in the original region; and derive the velocity field by Theorem 7.1.

Example 7.1 To picture the velocity field for a uniform flow around the "inside" of a corner, we can think of the fluid as moving down and to the right in the first quadrant of the XY-plane.

The mapping $z = w^2$ transforms the quadrant into the upper half plane where $h(z) = kz$ gives a complex potential for some real positive

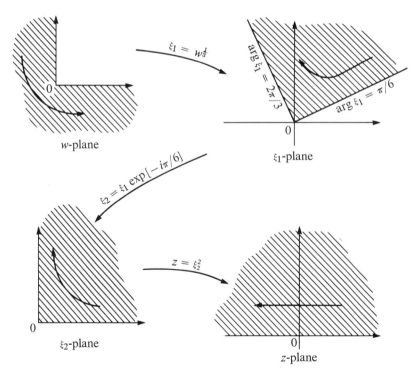

Figure 7.1

constant k. Then $H(w) = kw^2$ is a complex potential in the first quadrant, and $F(w) = 2k\overline{w}$ gives the velocity field there.

The equipotentials are the half arms of the hyperbolas $u^2 - v^2 = c$, and the streamlines are arms of the hyperbolas $uv = d$. (See problem 1.32.)

Example 7.2 Now consider a uniform flow around the "outside" of a corner — that is, a fluid moving down and to the right through the second, third, and fourth quadrants. A sequence of mappings tracing direction of flow is schematically outlined in Fig. 7.1. In the z-plane we can take $h(z) = -kz$, where k is a positive constant. If we use the branch of $w^{1/3}$ defined by

$$w^{1/3} = r^{1/3} \exp\left(\frac{i\theta}{3}\right), \quad r > 0, 0 < \theta < 2\pi$$

a complex potential for the flow is

$$H(w) = w^{2/3} \exp\left(\frac{-i\pi}{3}\right).$$

PROBLEMS

7.1 Find the complex potential, equipotentials, and streamlines of a uniform flow to the right in the region $D = \{z : |z| > 1, \text{Im } z > 0\}$. This represents a uniform flow past a cylindrical obstruction.

7.2 Find a complex potential and velocity field for a uniform flow to the right in the region D of Fig. 7.2. This represents flow down a channel with an obstruction splitting the flow. (Use problem 6.10.)

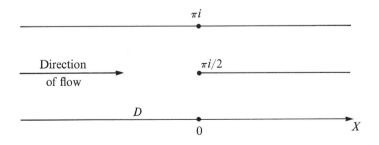

Figure 7.2

Section 7.2   Green's Functions and Poisson's Equation

Let D be a simply connected domain in the plane whose boundary C is a closed path. If F is a real-valued function defined in D, then the linear, second-order partial differential equation

$$\nabla^2 u = -F \quad \text{in } D$$

is called *Poisson's equation* (in two dimensions). It contains Laplace's equation as a special case.

If, in addition, we prescribe a real-valued function f on C, we can try to find a solution u in D to this boundary value problem:

$$\nabla^2 u = -F \quad \text{in } D$$
$$u = f \quad \text{on } C$$

In the special case where $F = 0$, this is the Dirichlet problem studied earlier.

This boundary value problem is a reasonable mathematical model for some physical problems of electrostatic potential and steady-state temperature distributions. The solution to the problem is intimately related to the Green's function for D studied in Chapter 6. We shall make enough additional assumptions on D, C, F, and f in order to be able to derive a solution to the problem.

We assume throughout that the boundary, C, of D is a closed path possessing at each point a unit vector $\mathbf{N}$ perpendicular to C and pointing out from D. We call $\mathbf{N}$ the outward unit normal at each point of C.

Lemma 7.1 (Green's Identities) Hypotheses: u and v are real-valued functions which have continuous first partial derivatives in $D \cup C$ and continuous second partial derivatives in D.

Conclusions:

C1
$$\iint_D u \nabla^2 v \, dx \, dy = \int_C u \nabla v \cdot \mathbf{N} \, ds - \iint_D (\nabla u \cdot \nabla v) \, dx \, dy$$

(Green's first identity)

C2 $$\iint\limits_{D} (u\nabla^2 v - v\nabla^2 u)\, dx\, dy = \int\limits_{C} (u\nabla v \cdot \mathbf{N} - v\nabla u \cdot \mathbf{N})\, ds$$

(Green's second identity)

where s denotes distance along C measured in the positive direction.

Outline of Proof: Since $\nabla \cdot (u\nabla v) = \nabla u \cdot \nabla v + u\nabla \cdot \nabla v = \nabla u \cdot \nabla v + u\nabla^2 v$,

$$\iint\limits_{D} u\nabla^2 v\, dx\, dy = \iint\limits_{D} \nabla \cdot (u\nabla v)\, dx\, dy - \iint\limits_{D} \nabla u \cdot \nabla v\, dx\, dy.$$

The divergence form of Green's theorem in the plane implies that

$$\iint\limits_{D} \nabla \cdot (u\nabla v)\, dx\, dy = \int\limits_{C} (u\nabla v) \cdot \mathbf{N}\, ds = \int\limits_{C} u\nabla v \cdot \mathbf{N}\, ds.$$

Thus

$$\iint\limits_{D} u\nabla^2 v\, dx\, dy = \int\limits_{C} u\nabla v \cdot \mathbf{N}\, ds - \iint\limits_{D} \nabla u \cdot \nabla v\, dx\, dy \qquad (7.1)$$

which is Green's first identity.

Similarly,

$$\iint\limits_{D} v\nabla^2 u\, dx\, dy = \int\limits_{C} v\nabla u \cdot \mathbf{N}\, ds - \iint\limits_{D} \nabla u \cdot \nabla v\, dx\, dy \qquad (7.2)$$

If we subtract Equation (7.2) from Equation (7.1) we obtain Green's second identity:

$$\iint\limits_{D} (u\nabla^2 v - v\nabla^2 u)\, dx\, dy = \int\limits_{C} (u\nabla v \cdot \mathbf{N} - v\nabla u \cdot \mathbf{N})\, ds \qquad (7.3)$$

THEOREM 7.2 Hypotheses:
H1 D is a simply connected domain, the closed path C is its boundary, and $\mathbf{N}$ is an outward unit normal at each point of C.
H2 $u(z)$ is real-valued, having continuous second partial derivatives in D and continuous first partial derivatives in $D \cup C$.

H3 $F(z)$ is real-valued and continuous in $D \cup C$.
H4 $u(z)$ satisfies the equations

$$\nabla^2 u = -F(z) \quad \text{in } D$$

$$u = 0 \qquad \text{on } C$$

Conclusion: For any z_0 in D,

$$u(z_0) = \frac{1}{2\pi} \iint\limits_D g(z,z_0)\, F(z)\, dx\, dy$$

where $g(z,z_0)$ is the Green's function for D with singularity at z_0.

Outline of Proof: For a fixed z_0 in D, choose a number $r > 0$ so small that $\triangle = \{z : |z - z_0| < r\}$ and $\Gamma = \{z : |z - z_0| = r\}$ lie in D. We shall

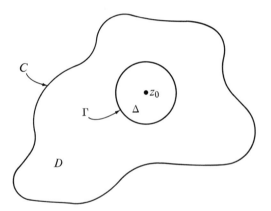

Figure 7.3

assume without proof that we can extend Lemma 7.1 to apply to the domain consisting of points of D outside Γ whose boundary is $C \cup \Gamma$. Call this domain E and its boundary S. Using Equation (7.3) in this case, we have

$$\iint\limits_E [g(z,z_0)\nabla^2 u - u\nabla^2 g(z,z_0)]\, dx\, dy$$

$$= \int\limits_S [g(z,z_0)\nabla u \cdot \mathbf{N} - u\nabla g(z,z_0) \cdot \mathbf{N}]\, ds$$

$$= \int_C [g(z,z_0)\nabla u \cdot \mathbf{N} - u\nabla g(z,z_0)\cdot \mathbf{N}] \ ds$$

$$- \int_\Gamma [g(z,z_0)\nabla u \cdot \mathbf{N} - u\nabla g(z,z_0)\cdot \mathbf{N}] \ ds,$$

where at each point of S, $\mathbf{N}$ is the outward unit normal to E.

Since $\nabla^2 g(z,z_0) = 0$ in E, $g(z,z_0) = 0$ on C, and $u = 0$ on C, we have

$$- \iint_E g(z,z_0) \ F(z) \ dx \ dy = - \int_\Gamma [g(z,z_0)\nabla u \cdot \mathbf{N} - u\nabla g(z,z_0)\cdot \mathbf{N}] \ ds. \quad (7.4)$$

If we let $r \to 0$, the left side of (7.4) approaches $-\iint_D g(z,z_0) \ F(z) \ dx \ dy$. From the definition of $g(z,z_0)$ we know that on Γ $g(z,z_0) = -\log |z - z_0| + T(|z - z_0|) = -\log r + T(r)$, where $T(r)$ is twice differentiable and $\lim_{r\to 0} T(r) = 0$. Also, on Γ the outward unit normal, $\mathbf{N}$, to E is given by $(z - z_0)/r$. Thus

$$\int_\Gamma g(z,z_0)\nabla u \cdot \mathbf{N} \ ds = 2\pi \int_0^{2\pi} [-\log r + T(r)]\nabla u \cdot \mathbf{N}(r \ d\theta)$$

$$= -2\pi \int_0^{2\pi} (r \log r)\nabla u \cdot \mathbf{N} \ d\theta + 2\pi \int_0^{2\pi} r \ T(r)\nabla u \cdot \mathbf{N} \ d\theta.$$

As $r \to 0$ both these integrals approach 0, since $\lim_{r\to 0} r \log r = 0$, and $\lim_{r\to 0} r \ T(r) = 0$, and $\nabla u \cdot \mathbf{N}$ is continuous.

Now on Γ,

$$\nabla g(z,z_0) = -\nabla \log r + \nabla T(r),$$

and

$$\nabla g(z,z_0) \cdot \mathbf{N} = -(\nabla \log r)\cdot \mathbf{N} + \nabla T(r)\cdot \mathbf{N}$$

$$= -\left(\frac{1}{r}\right) + \nabla T(r)\cdot \mathbf{N},$$

where $\nabla T(r)\cdot \mathbf{N}$ remains bounded as $r \to 0$. Then

$$-\int_{\Gamma} u \nabla g(z,z_0) \cdot \mathbf{N} \, ds = -\int_{0}^{2\pi} u\left[-\left(\frac{1}{r}\right) + \nabla T(r) \cdot \mathbf{N}\right](2\pi r \, d\theta)$$

$$= 2\pi \int_{0}^{2\pi} u(z_0 + re^{i\theta}) \, d\theta - 2\pi r \int_{0}^{2\pi} \nabla T(r) \cdot \mathbf{N} \, d\theta,$$

and as $r \to 0$ we obtain as a limit $2\pi u(z_0)$. Interpreting this result in (7.4) we have

$$\iint_{D} g(z,z_0) \, F(z) \, dx \, dy = 2\pi u(z_0).$$

Using Theorem 7.2 we can obtain another perspective on the Dirichlet problems considered in Chapter 6.

Corollary 7.1   Hypotheses:
H1, H2   as in Theorem 7.2.
H3 $f(z)$ is real-valued with continuous first partial derivatives on C.
H4 $u(z)$ satisfies the equations

$$\nabla^2 u = 0 \qquad \text{in } D$$

$$u = f(z) \quad \text{on } C$$

Conclusion: For any z_0 in D

$$u(z_0) = \frac{-1}{2\pi} \int_{C} f(z) \nabla g(z,z_0) \cdot \mathbf{N} \, ds$$

where $g(z,z_0)$ is the Green's function for D with singularity at z_0.

Outline of Proof: Note that we have imposed stronger hypotheses on the boundary function $f(z)$ than we did in Chapter 6. These hypotheses are not necessary for the present conclusion but only for the present method of proof based on Theorem 7.2.

For a fixed z_0 in D let $q(z)$ be any real-valued function such that:
(1) q has continuous second partial derivatives in D;
(2) q has continuous first partial derivatives on C;
(3) $q(z_0) = 0$;
(4) $q(z) = f(z)$ on C.

It is not obvious that such a function $q(z)$ need exist, but we assert that there is such a function and omit the details. We define $w(z) = u(z) - q(z)$ in $D \cup C$, so that $w(z)$ satisfies the equations

$$\nabla^2 w = -\nabla^2 q \quad \text{in } D$$

$$w = 0 \qquad \text{on } C.$$

By Theorem 7.2,

$$w(z_0) = \frac{1}{2\pi} \iint_D g(z,z_0)\nabla^2 q(z) \, dx \, dy.$$

The following steps are taken formally; to justify them requires an argument like the previous one where we "cut out" a small disk about z_0 and then examine limits as the radius of this disk shrinks to zero.

We write

$$g(z,z_0)\nabla^2 q(z) = \nabla \cdot [g(z,z_0)\nabla q(z) - q(z)\nabla g(z,z_0)] + q(z)\nabla^2 g(z,z_0),$$

so that

$w(z_0)$

$$= \frac{1}{2\pi}\left\{ \iint_D \nabla \cdot [g(z,z_0)\nabla q - q\nabla g(z,z_0)] \, dx \, dy + \iint_D q\nabla^2 g(z,z_0) \, dx \, dy\right\}$$

$$= \frac{1}{2\pi}\int_C [g(z,z_0)\nabla q \cdot \mathbf{N} - q\nabla g(z,z_0) \cdot \mathbf{N}] \, ds + \frac{1}{2\pi}\iint_D q\nabla^2 g(z,z_0) \, dx \, dy.$$

By an argument similar to that for Theorem 7.2, we can show that $\frac{1}{2\pi} \iint_D q\nabla^2 g(z,z_0) \, dx \, dy = q(z_0) = 0$. This implies that

$$w(z_0) = -\frac{1}{2\pi}\int_C q(z)\nabla g(z,z_0) \cdot \mathbf{N} \, ds$$

$$= -\frac{1}{2\pi}\int_C f(z)\nabla g(z,z_0) \cdot \mathbf{N} \, ds.$$

Taken together, Theorem 7.2 and Corollary 7.1 imply this: If there is a solution $u(z)$ to the problem

$$\nabla^2 u = -F(z) \quad \text{in} \quad D$$

$$u = f(z) \qquad \text{on} \quad C$$

which has continuous second partial derivatives in D and continuous first partial derivatives on C, then it must be this function for $z_0 \in D$:

$$u(z_0) = \frac{1}{2\pi} \iint_D g(z,z_0) F(z) \, dx \, dy - \frac{1}{2\pi} \int_C f(z) \nabla g(z,z_0) \cdot \mathbf{N} \, ds.$$

Since the Green's function $g(z,z_0)$ for D with singularity at z_0 does not change with the functions F, f, this formula provides an important means of estimating how solutions to the boundary value problems might be altered by various changes in the functions F and f.

Since Green's theorem and the vector operations we have used can be extended to higher dimensions, the basic form and outline of the arguments above can also be extended. What is lost in higher dimensions is a broadly applicable technique for finding Green's functions to compare with the conformal mapping technique of Chapter 6 which is available for two dimensions.

PROBLEMS

7.3 If $D = \{z : |z| < 1\}$ and $C = \{z : |z| = 1\}$, while $z = se^{i\phi}$, $z_0 = re^{i\theta}$,

(a)   write $g(z,z_0)$ in polar coordinates;

(b)   if $\mathbf{N}$ is the outward unit normal at each point $z = e^{i\phi}$ of C, show that

$$\nabla g(z,z_0) \cdot \mathbf{N} = \frac{1 - r^2}{1 + r^2 - 2r \cos(\theta - \phi)}$$

(c)   Show that

$$\nabla g(z,z_0) \cdot \mathbf{N} = \text{Re}\left(\frac{e^{i\phi} + re^{i\theta}}{e^{i\phi} - re^{i\theta}}\right).$$

7.4 Let $D = \{z : |z| < 1\}$, $C = \{z : |z| = 1\}$, and suppose $f(z)$ is defined with continuous first partial derivatives on C. Suppose $u(z)$ has continuous second partial derivatives in D and continuous first partial derivatives on C and solves the problem

$$\nabla^2 u = 0 \qquad \text{in } D$$

$$u = f(z) \quad \text{on } C.$$

Deduce that

$$u(re^{i\theta}) = \frac{1}{2\pi} \int_0^{2\pi} \frac{f(e^{i\phi})(1 - r^2)}{1 + r^2 - 2r\cos(\theta - \phi)} \, d\phi \qquad (7.5)$$

for $z_0 = re^{i\theta} \in D$.

Equation (7.5) is the *Poisson integral formula* for the open unit disk. It can be derived from the Poisson integral formula for the upper half plane (Theorem 6.5) by a conformal mapping argument.

7.5   In problem 7.4:
   (a)   show that for each $z_0 = re^{i\theta} \in D$

$$\min_C f(z) \le u(re^{i\theta}) \le \max_C f(z);$$

   (b)   show that if $f(z) = k$, a constant, on C, then $u(z_0) = k$ for all $z_0 \in D$. (*Hint:* Recall problem 3.23.)

8

Analytic Continuation

In all our dealings with analytic functions thus far we have thought consistently in terms of an analytic function as being a function "analytic in some domain." There are a variety of advantages in arriving at a broader, more general concept of "analytic function" which of course includes the information we already have.

At each point z_0 where a given function F is analytic, we can obtain a Taylor's series representation for F converging in an open disk centered at z_0. It is natural to think of this series and its disk of convergence* as a "local representation" of F near z_0. The new approach we wish to take is to think about an analytic function as the collection of all its possible local representations.

---

*Our convention from Chapter 4 is that a "disk of convergence" is always an open disk.

Example 8.1   Suppose $D_1 = \{z: |z| < 1\}$ and $D_2 = \{z: |z - i| < 2^{1/2}\}$. The series $f_1(z) = \sum_{n=0}^{\infty} z^n$ has D_1 as its disk of convergence, while

$$f_2(z) = \left[\frac{1}{(1 - i)}\right] \sum_{n=0}^{\infty} \left[\frac{(z - i)}{(1 - i)}\right]^n$$

has D_2 as its disk of convergence. Each of the pairs, (f_1, D_1) and (f_2, D_2), is a local representation for the function $F(z) = 1/(1 - z)$. Note that the disks D_1 and D_2 intersect, and at each point z_1 in $D_1 \cap D_2$ we would have $f_1(z_1) = f_2(z_1)$.

If a function F is analytic in a given domain, we know how to get its local representations by power series. Suppose, however, we are given a power series f_1 with disk of convergence D_1. We would like to be able to think about the existence of some function F for which the pair (f_1, D_1) is a local representation. In Example 8.1 we could write a formula for F, but we will find it useful to think of such an F even when we have no general formula for it.

Section 8.1   General Analytic Functions

Now it is time to make more precise statements. Example 8.1 will serve as a concrete model for our language.

Definition   Suppose f_1 is a power series with disk of convergence D_1 and f_2 is a power series with disk of convergence D_2. If D_1 and D_2 intersect, and if $f_1(z) = f_2(z)$ for each z in $D_1 \cap D_2$, we say f_2 is the *direct analytic continuation of f_1 to D_2* (or f_1 is the *direct analytic continuation of f_2 to D_1*.)

From the facts we have studied for power series, we can see that if (f_1, D_1) has a direct analytic continuation to D_2, then it has only one such continuation.

The process of direct analytic continuation may be continued. If each two adjacent entries in the finite sequence

$$(f_1, D_1), (f_2, D_2), \ldots, (f_n, D_n)$$

are direct analytic continuations of each other, we say that there is a *direct analytic continuation chain* from (f_1, D_1) to (f_n, D_n).

Definition   The *general analytic function F containing* (f_1, D_1) is the set of all possible pairs (f_n, D_n), $n = 1, 2, \ldots$, such that:

(1)   for each n, f_n is a power series with disk of convergence D_n;

(2)   for each n and m, there is a direct analytic continuation chain from (f_n, D_n) to (f_m, D_m). If z_n is the center of D_n, we call each pair (f_n, D_n) an *analytic element of F at* z_n.

Suppose f is a power series with disk of convergence D. Without giving proofs, we state an important fact about general analytic functions.

THEOREM 8.1   Hypothesis: f is a power series and D is its disk of convergence.

Conclusion: There is exactly one general analytic function F which contains the analytic element (f, D).

In Example 8.1, $F(z) = 1/(1 - z)$ is the general analytic function containing the analytic element (f_1, D_1), where $f_1(z) = \sum_{n=0}^{\infty} z^n$ and $D_1 = \{z : |z| < 1\}$. It is possible for a general analytic function to consist of a single analytic element. Such a case arises for the power series $f(z) = \sum_{n=0}^{\infty} z^{2^n}$ and its disk of convergence $D_1 = \{z : |z| < 1\}$. (The reader who wishes to see a proof of this can look in [Copson, pp. 85–88].)

Section 8.2   Analytic Continuation Along a Path

Before we can illustrate the utility of the concept of general analytic function, we must extend slightly the notion of direct analytic continuation.

Suppose f is a power series with disk of convergence D centered at z_0, g is a power series with disk of convergence E centered at z_1, and $C : \{z(t) : 0 \leq t \leq 1\}$ is a path from $z_0 = z(0)$ to $z_1 = z(1)$.

Definition   We call (g, E) an *analytic continuation of* (f, D) *along* C if there is a finite set of values $0 = t_0 < t_1 < \cdots < t_n = 1$ and a direct analytic continuation chain (f_0, D_0), (f_1, D_1), $\ldots$, (f_n, D_n) such that D_j is centered at $z(t_j)$, $j = 0, 1, \ldots, n$, and $f_n = g$ in $D_n \cap E$.

If we know a function F is analytic in some domain G with local representation (f, D) at z_0 in G, we expect any analytic continuation along a path from z_0 in G to any other point z_1 of G to yield precisely the local representation of F at z_1. And this is indeed the case.

On the other hand, suppose f is a power series with disk of convergence D centered at z_0, C_1 and C_2 are distinct paths from z_0 to z_1. If (f,D) can be continued analytically along both C_1 and C_2 to the point z_1, the resulting continuations at z_1 need not be the same. (Example 8.2 will illustrate such a case.) A famous theorem, known as the *Monodromy Theorem* (from the Greek for "one way"), gives conditions which are sufficient for us to avoid this problem.

THEOREM 8.2   Hypotheses:
H1 G is a simply connected domain and z_0 lies in G.
H2 f is a power series with disk of convergence D centered at z_0, $D \subset G$.
H3 (f,D) can be analytically continued along every path in G starting at z_0.

Conclusions:
C1   Analytic continuations of (f,D) along all paths in G starting at z_0 and ending at a point z_1 of G must yield the same analytic element at z_1.
C2   There exists a function F analytic in G such that $F(z) = f(z)$ for z in D.*
The Monodromy Theorem has an immediate and useful consequence.

Corollary 8.1   Hypotheses:
H1 f is a power series with disk of convergence D centered at z_0.
H2 C_1 and C_2 are paths from z_0 to z_1, intersecting only at z_0 and z_1, along which (f,D) has different analytic continuations to z_1.

Conclusion: The general analytic function containing (f,D) must have at least one singularity in the domain bounded by $C_1 \cup C_2$.

Example 8.2   Let $f(z) = \sum_{n=1}^{\infty} [(-1)^{n+1}/n](z-1)^n$ in the disk $D = \{z : |z-1| < 1\}$. The pair (f,D) is a local representation for Log z at $z = 1$.

For each path C, not passing through 0, from the point 1 to any $w \neq 0$ define

$$F(w,C) = \int_{1 \atop C}^{w} \frac{dz}{z}.$$

---

*If (f_n, D_n) is any analytic element of F and z is in D_n, we denote by $F(z)$ the unique complex number $w = f_n(z)$. In the absence of simple connectivity of G, the symbol "$F(z)$" is ambiguous, since there might be many analytic elements of F whose disks contain z. Thus F might attain many distinct values at a. The notion of a *Riemann surface* is required to provide an unambiguous meaning to $F(z)$ in such cases. Example 8.2 brings out certain features of this difficulty.

If w is real and positive and C is a segment of the positive real axis, then $F(w,C)$ produces the usual natural logarithm of w. The number $F(w,C)$ is the value at w of the analytic continuation to w along C of (f,D).

Let C_1 be the path from 1 to i along the unit circle $\{z:|z| = 1\}$ in the positive direction. Then

$$F(i,C_1) = \int_{\substack{1 \\ C_1}}^{i} \frac{dz}{z} = i \int_0^{\pi/2} d\theta = \frac{\pi i}{2}.$$

Now let C_2 be the path from 1 to i along the unit circle in the negative direction. Then

$$F(i,C_2) = \int_{\substack{1 \\ C_2}}^{i} \frac{dz}{z} = i \int_0^{-3\pi/2} d\theta = \frac{-3\pi i}{2}.$$

Corollary 8.1 implies that the general analytic function containing (f,D) has a singularity somewhere inside the unit circle. (From what we know of log z, we can be sure the singularity is $z = 0$.)

The connection between the present discussion and the treatment of logarithms in Chapter 2 is exactly this: Were we to cut the plane along a ray from $z = 0$ and prohibit our paths C from crossing this ray, the value $F(w,C)$ for any w not on the ray would be independent of the path C. Thus $F(w,C)$ would be a way of defining Log w analytically for all w not on the ray.

By cutting the plane along a ray from $z = 0$, we are left with a simply connected domain. The uniqueness of analytic continuation within this domain thus is a consequence of the Monodromy Theorem.

Section 8.3 The Law of Permanence of Functional Equations

Our discussion of the idea of analytic continuation to this point has offered little in the way of concrete techniques for its application. The mechanical difficulties in carrying out any analytic continuation directly by manipulation of power series are very formidable. Yet the idea of analytic continuation permits applications of great power and finesse. As a means of motivating a broad and powerful principle we present a simple example.

Example 8.3   We wish to find a real-valued solution $y(x)$ to the differential equation problem

$$(x - 1)y'(x) + y(x) = 0$$

$$y(0) = 1$$

which is valid for as many values of x as possible.

Let us replace x by the complex variable z and try to solve the problem

$$(z - 1)y'(z) + y(z) = 0$$

$$y(0) = 1$$

by a method called the method of power series. That is, assume

$$y(z) = \sum_{n=0}^{\infty} a_n z^n$$

is a solution to the problem and try to determine $a_0, a_1, \ldots, a_n, \ldots$. If
this method results in a power series solution with a disk of convergence
centered at $z = 0$, we can hope that the general analytic function having
this local representation also solves the problem. Interpreting this
information on the real axis should provide the solution we want.

If $y(0) = 1$, then $a_0 = 1$. Also

$$(z - 1) \sum_{n=0}^{\infty} n \, a_n z^{n-1} + \sum_{n=0}^{\infty} a_n z^n = 0.$$

A rearrangement of this equation gives

$$\sum_{n=0}^{\infty} [(n \, a_n + 1) - (n + 1)a_{n+1}]z^n = 0.$$

This suggests that

$$a_{n+1} = \frac{1}{n + 1} + \left(\frac{n}{n + 1}\right)a_n, \quad \text{for} \quad n = 0, 1, 2, \ldots.$$

This recursion relation shows us that

$$a_{n+1} = a_0 = 1 \quad \text{for} \quad n = 0, 1, 2, \ldots.$$

Thus $y(z) = \sum_{n=0}^{\infty} z^n$ seems to solve our problem, at least for all z in $D = \{z : |z| < 1\}$. But we know (y, D) is a local representation for $F(z) = 1/(1 - z)$. It is easy to see that F solves the problem for all z different from 1. Thus on the real axis the problem has solution $F(x) = 1/(1 - x)$ for all real x except $x = 1$.

This is a simple problem, and it can be solved more easily than we have indicated. But the solution we have outlined illustrates the features of a general theorem called the *law of permanence of functional equations*. The version we state here is far from the most general.

THEOREM 8.3   Hypotheses:
H1 $f_1, f_2, \ldots, f_n$ are power series with a common disk of convergence D.
H2 $G(z, z_1, z_2, \ldots, z_n)$ is a complex-valued function of $n + 1$ variables which is analytic in each variable separately for all complex values.
H3 for each z in D

$$G(z, f_1(z), f_2(z), \ldots, f_n(z)) = 0.$$

H4 $(f_1, D), (f_2, D), \ldots, (f_n, D)$ can be analytically continued along a path C to $(g_1, E), (g_2, E), \ldots, (g_n, E)$, respectively.

Conclusion: for each z in E

$$G(z, g_1(z), g_2(z), \ldots, g_n(z)) = 0.$$

We might loosely paraphrase the theorem in this way: If $f_1, f_2, \ldots, f_n$ are analytic in a common disk and obey a functional relation in this disk, then their analytic continuations also obey the same functional relation.

Since any differential equation imposes a functional relation on any solution and some of its derivatives, if we can obtain a power series solution valid in a disk, its analytic continuations should provide a solution valid in larger regions.

Example 8.3 is the special case of Theorem 8.3 which arises when $n = 2$, $f_1(z) = y(z)$, $f_2(z) = y'(z)$, $D = \{z : |z| < 1\}$, and $G(z, z_1, z_2) = (z - 1)z_2 + z_1$.

Many of the elementary functions we studied in Chapter 2 can be characterized in terms of functional equations. (Recall Problems 2.7, 2.8.)

Example 8.4 Suppose $f(z)$ is a function analytic in $D = \{z: |z| < 1\}$ such that $f(0) = 1$, and for $-1 < x < 1$, f satisfies the relation

$$f(2x) = 2[f(x)]^2 - 1.$$

The Monodromy Theorem and the law of permanence imply that f must satisfy the same relation throughout D. For any w, $|w| < 2$, we see that

$$f(w) = 2\left[f\left(\frac{w}{2}\right) \right]^2 - 1,$$

and, since f is analytic in D, we can see that f must be analytic in $\{w: |w| < 2\}$. In the same way we can see that f is analytic in $\{w: |w| < 4\}$. In fact for any positive integer n we can argue that f is analytic in $\{w: |w| < 2^n\}$. Repeated use of the Monodromy Theorem and the law of permanence shows the original relation must persist in the larger disks. The result is that f is analytic for all z. Can you guess what function f might be? Is it unique?

PROBLEMS

8.1 If $w \neq 0$ and C is any path from 1 to w not passing through 0, define

$$G(w,C) = 1 + \int_{\substack{1 \\ C}}^{w} \frac{dz}{2\, z^{1/2}}.$$

    (a) If w is real and positive, and C is a segment of the positive real axis, what is $G(w,C)$?

    (b) By choosing C in two different ways determine two values for $G(-1,C)$. Why is it that $G(-1,C)$ has no more than two distinct values? (*Hint:* Recall that $z^{1/2} = \exp[(\frac{1}{2}) \log z]$.)

8.2 Suppose $f(z)$ is to satisfy this equation for all z:

$$f(z) = e^z + \int_{0}^{z} f(w) \exp(w - z)\, dw.$$

    (a) Write a differential equation whose solution satisfies the same equation as f.

    (b) Obtain a solution for f by solving a problem involving this differential equation.

8.3   If $g(z)$ is a given entire function and $f(z)$ satisfies the equation:

$$f(z) = g(z) + \int_0^z f(w) \exp(w - z)\, dw,$$

(a)   what differential equation problem must f solve?

(b)   if the function f is analytic at any point $z = a$, explain why, in fact, f must be analytic for all z.

8.4   Suppose D and E are intersecting open disks and f and g are power series with disks of convergence D and E, respectively.

(a)   If g is the direct analytic continuation of f to E, explain why g' is the direct analytic continuation of f' to E.

(b)   If f can be analytically continued along a path C starting from the center of D, explain why f' can be continued along C also.

(c)   If F is the general analytic function containing (f, D), what name would you give to the general analytic function containing (f', D)?

Section 8.4   Functions Defined by Integrals

Many functions of interest in applied mathematics are complex-valued functions defined by a definite integral with respect to a real parameter. The Fourier transform discussed briefly in Chapter 5 is a specific example. Suppose $F(z) = \int_a^b f(z, t)\, dt$, where $f(z, t)$ is continuous for $a \le t \le b$ and all z in a simply connected domain D. If $F(z)$ is analytic in D, and if we can actually evaluate $F(z)$ at points of some subset of D, Theorems 8.2 and 8.3 may help us evaluate $F(z)$ in all of D. Even if we can't find a formula for $F(z)$ in D, we might be able to use the theorems to obtain useful properties for $F(z)$ throughout D.

It is helpful to have a statement giving some conditions for the integral $F(z)$ to be analytic in D.

THEOREM 8.4   Hypotheses:

H1 D is a simply connected domain and $a \le t \le b$ is a real interval of finite length.

H2 $f(z, t)$ is defined and continuous for all z in D and $a \le t \le b$, and for each fixed t, $a \le t \le b$, $f(z, t)$ is analytic in D.

Conclusions:

C1 $F(z) = \int_a^b f(z,t) \, dt$ is analytic in D.

C2   For each z in D

$$F'(z) = \int_a^b \frac{\partial}{\partial z} f(z,t) \, dt,$$

where $(\partial/\partial z)f(z,t)$ denotes the derivative with respect to z of $f(z,t)$ for each fixed value of t.

Example 8.5   If $f(z,t) = e^{zt} \cos t$, the hypotheses of Theorem 8.4 are satisfied for all real t and all finite z. Thus $F(z) = \int_0^\pi e^{zt} \cos t \, dt$ is analytic for all finite z. For each fixed real x we can use integration by parts to show that $F(x) = -x(e^{\pi x} + 1)/(x^2 + 1)$. An analytic continuation argument shows us that for all finite z, $F(z) = -z(e^{\pi z} + 1)/(z^2 + 1)$. (Is it obvious that $F(z)$ is analytic at $z = i, -i$?)

Like Fourier transforms, many of the integrals we might wish to examine are improper integrals. A result like Theorem 8.4 for integrals of the form $\int_0^\infty f(z,t) \, dt$ is available, but its hypotheses are beyond the scope of our discussion. A good reference for details and examples is [Copson, pp. 106–115].

Example 8.6   It can be proved that $F(z) = \int_0^\infty \exp(-zt) \, dt$ is analytic for all z such that Re $z > 0$. For real $x > 0$, it is easy to verify that $F(x) = 1/x$. The argument by analytic continuation yields the formula

$$F(z) = \int_0^\infty \exp(-zt) \, dt = \frac{1}{z}$$

for Re $z > 0$.

PROBLEMS

8.5   In Example 8.5 what are $F(i)$ and $F(-i)$?

8.6   Use the result of Example 8.5 to determine a formula for the function $G(z) = \int_0^\pi t\, e^{zt} \cos t\, dt$.

8.7   Let $F(z) = \int_0^{2\pi} dt/(1 + z \cos t)$.
   (a)   In what domain is $F(z)$ analytic?
   (b)   Determine $F(x)$ for all real x in this domain.
   (c)   What is the value of $F(z)$ for each z where it is analytic?

8.8   The *gamma function*, a well-known special function of applied mathematics, is defined by the improper integral

$$\Gamma(z) = \int_0^\infty e^{-t}\, t^{z-1}\, dt$$

and is known to be analytic in the domain $\{z : \operatorname{Re} z > 0\}$.
   (a)   Verify that, for all $x > 0$, $\Gamma(x + 1) = x\Gamma(x)$.
   (b)   Write out the argument showing that $\Gamma(z + 1) = z\Gamma(z)$ in the domain $\{z : \operatorname{Re} z > 0\}$.

8.9   Another important special function, the *beta function*, can be defined in terms of the gamma function:

$$B(x,y) = \frac{\Gamma(x)\Gamma(y)}{\Gamma(x + y)}, \qquad x > 0, \, y > 0.$$

The beta function also has an integral representation

$$B(x,y) = \int_0^\infty \frac{t^{x-1}\, dt}{(1 + t)^{x+y}}.$$

   (a)   Use this representation and Example 5.10 to evaluate $\Gamma(x)\Gamma(1 - x)$ for $0 < x < 1$.
   (b)   What property does (a) reveal for $\Gamma(z)$ in the domain $\{z : 0 < \operatorname{Re} z < 1\}$?

REFERENCES

Carrier, F., M. Krook, and C. E. Pearson. *Functions of a Complex Variable*. New York: McGraw-Hill, 1966.

Churchill, R. V. *Introduction to Complex Variables and Applications* (2nd ed.). New York: McGraw-Hill, 1960.

Copson, E. T. *The Theory of Functions of a Complex Variable*. London: Oxford University Press, 1935.

Cunningham, J. *Complex Variable Methods in Science and Technology*. New York: D. Van Nostrand, 1965.

Hille, E. *Analytic Function Theory*, Vol. I. New York: Ginn, 1959.

Kaplan, W. *Introduction to Analytic Functions*. Reading, Mass.: Addison-Wesley, 1966.

Kober, H. *Dictionary of Conformal Representations*. New York: Dover, 1952.

Nehari, Z. *Introduction to Complex Analysis* (rev. ed.). Boston: Allyn and Bacon, 1968.

Pennisi, L. L. *Elements of Complex Variables*. New York: Holt, Rinehart and Winston, 1963.

Spiegel, M. R. *Theory and Problems of Complex Variables* (Schaum's Outline Series). New York: Schaum Publishing, 1964.

ANSWERS TO PROBLEMS

Chapter 1

1.1 (a) $2i, -i, -i$
(b) $2 + 2i, -2 + 2i, 1/2 - i/2$
(c) $6 + 8i, -117 + 44i, 3/25 - (4/25)i$

1.2 (a) $z_1 = 2^{1/2}[\cos(\pi/4) + i\sin(\pi/4)]$,
$z_2 = 5^{1/2}[\cos(5.82) + i\sin(5.82)]$
(b) $z_1 + z_2 = 3, z_1z_2 = 3 + i = 10^{1/2}[\cos(.32) + i\sin(.32)]$,
$z_1/z_2 = 1/5 + (3/5)i = (.632)[\cos(.46) + i\sin(.46)]$
(c) (i) $2x - 4y = 3$
(ii) $32x^2 + 16xy + 20y^2 - 96x - 24y + 36 = 0$

1.4 z_1 and z_2 must be on the same ray drawn from the origin.

1.6 The set of points lying on the same side as b of the perpendicular bisector of the line segment joining a and b.

1.7 (a) $z = \cos(2\pi k/3) + i\sin(2\pi k/3), k = 0,1,2$.
(b) $z = \cos(\pi/5 + 2\pi k/5) + i\sin(\pi/5 + 2\pi k/5), k = 0,1,2,3,4$.
(c) $z = 3^{1/6}[\cos(\pi k/3) + i\sin(\pi k/3)], k = 0,1,2,3,4,5$.
(d) $z = (1/2)[-1 \pm i\sqrt{11}]$
(e) $z = 2[\cos(\pi/6 + 2\pi k/3) + i\sin(\pi/6 + 2\pi k/3)], k = 0,1,2$.

1.9 (c) $[1 + \cos x - \cos nx - \cos(n - 1)x]/(2 - 2\cos x)$

1.10 (c) and (d) are open; all are connected; (c) and (d) are domains; (c) is a simply connected domain.

1.17 (b), (c), and (f) are compact.

1.22 $f'(z_0) = \cos x_0 \cosh y_0 - i\sin x_0 \sinh y_0, f(x_0) = \sin x_0$,
$f'(x_0) = \cos x_0$.

1.29 No; no.

1.31 (a) $v(x,y) = x/(x^2 + y^2)$
(b) $v(x,y) = \cos x \sinh y$
(c) $v(x,y) = 2xy$

1.33 If $f(x + iy) = u(x,y) + iv(x,y), f'(0) = 0$.

Chapter 2

2.3 $z = 2k\pi i$, k any integer.

2.5 (a) $z = \log 5 + i(\pi/2 + 2k\pi)$, k any integer.
(b) $z = (1/2)\log 2 + i(\pi/4 + 2k\pi)$, k any integer.
(c) $z = x + iy$, $y > 0$.

2.10 (a) $z = \pm \cos^{-1}(4/\sqrt{31}) + 2k\pi - (i/2)\log 31$, k any integer.
(b) $z = \pm \cos^{-1}(2/3) + (2k + 1)\pi - i \log 3$, k any integer.
(c) $z = k\pi + (i/2)\log 5$, k any integer.

2.12 (a) $z = k\pi i$, k any integer.
(b) $z = (2k + 1)\pi i/2$, k any integer.
(c) $z = \log(2 \pm \sqrt{3}) + i(\pi/2 + 2k\pi)$, k any integer.
(d) $z = \log(4 \pm \sqrt{15}) + 2k\pi i$, k any integer.

2.17 $x = \log[y + (y^2 + 1)^{1/2}]$.

2.18 (a) $z = \text{Log } 17 + 2k\pi i$, k any integer.
(b) $z = (1/2)\text{Log } 3 + (2k + 1)\pi i/2$, k any integer.

2.19 $(1/2)\text{Log } 2 - i\pi/4$.

2.20 (a) $\pm \exp(i\pi/4)$.
(b) $2^{1/8} \exp[i(\pi/16 + k\pi/2)]$, $k = 0,1,2,3$.
(c) $2 \exp(2k\pi i/3)$, $k = 0,1,2$.
(d) $\exp(-\pi/2 - 2k\pi)$, k any integer.

Chapter 3

3.1 0.

3.3 Yes.

3.6 $32\sqrt{2}$.

3.8 $1 + 2i/3$

3.9 0 if k is even, -2 if k is odd.

3.10 $a \text{ Log } a - a + 1$.

3.11 $(2/3)(a^{3/2} - 1)$.

3.12 $2\pi i$ if $n = -1$, 0 otherwise.

3.13 0 for all n if a is outside C; 0 for all $n \neq 1$ if a is inside C; $2\pi i$ if $n = 1$ and a is inside C.

3.14 (a) 0
    (b) $-\pi i$
    (c) πi

3.15 (a) 0
    (b) $-\pi i$
    (c) πi

3.16 No; let $f(z) = \text{Re } z$.

3.17 (a) $-2\pi i$
    (b) $2\pi i(e^2 - e)$
    (c) $2\pi i$
    (d) 0 if C does not enclose 2; $2\pi i$ if C encloses 2.
    (e) 0 if $m \neq n$; $2\pi i$ if $m = n$.

3.21 80/27

3.22 At the point(s) of W furthest to the right of the imaginary axis.

3.26 2 at $z = \pm 1$; 0 at $z = \pm i$.

Chapter 4

4.1 (a) $\rho = \infty$; converges for all finite z.
    (b) $\rho = \infty$; converges for all finite z.
    (c) $\rho = 1$; $|z - 2| < 1$.
    (d) $\rho = 0$; converges only for $z = 1$.

4.2 (a) $|2z + 3| < 3$.
    (b) $|z + 1| < \sqrt{2}$.
    (c) $|3z + 2| < 1$.

4.4 (a) $|z| < 1$; $1 - \text{Log}(1 - z)$.
    (b) $|z| < 1$; $z(1 + z)/(1 - z)^3$.
    (c) $|z| < 1$; $-\int_0^z (1/z)\text{Log}(1 - z)dz$.
    (d) $|3z + 2| < 1$; $(3z + 2)/(3z + 1)^2$.

4.6 $\{z: \text{Im } z > 0\}$; $k - \text{Log}[1 - \exp(2\pi iz)]$, where
$$k = \sum_{n=0}^{\infty} \exp[-2\pi(n + 1)]/(n + 1).$$

4.8 $\sum_{n=0}^{\infty} z^{n+1}/(n + 1)$; $|z| < 1$.

4.9 (a) $\sum_{n=0}^{\infty} (-1)^n z^{2n}$

  (b) $\sum_{n=0}^{\infty} (-1)^n z^{2n+1}/(2n+1)$

4.10 (a) $z - z^3/3! + z^5/5! - z^7/7!$
  (b) $1 - z^2/2! + z^4/4! - z^6/6!$
  (c) $z - z^2/2 + z^3/3 - z^4/4$
  (d) $z + z^2/2 + z^3/3 - z^5/120$
  (e) $z + z^2 + z^3/3 - z^5/30$
  (f) $e + ez + ez^2 + (2/3)ez^3$
  (g) $-2z - 2z^3 - 2z^5 - 2z^7$

4.11 $-\sum_{n=0}^{\infty} z^{2n+1}/(2n+1)!$

4.12 $-\sum_{n=0}^{\infty} (z+1)^n$

4.13 $-1 - 2\sum_{n=1}^{\infty} z^n$

4.14 (a) $1/4z + \sum_{n=1}^{\infty} (-1)^n z^{2n-1}/2^{2n+2}$

  (b) $\sum_{n=0}^{\infty} (-1)^n 2^{2n} z^{-2n-3}$
  (c) $-1/6(z - 2i) - i/6 + (5/36)(z - 2i) + (i/9)(z - 2i)^2 + \dots$
  (d) $1/(z - 2i)^3 - 6i/(z - 2i)^4 - 30/(z - 2i)^5$
     $+ 216i/(z - 2i)^6 - \dots$

4.15 (a) $1/2 - (1/4)z - (3/8)z^2 + (3/16)z^3 + (13/32)z^4 - (13/64)z^5 \dots$
  (b) $1/z^3 - 2/z^4 + 3/z^5 - 6/z^6 + 13/z^7 \dots$
  (c) $1/z^3 - 2/z^4 - 5/z^5 + 10/z^6 - 11/z^7 \dots$
  (d) $1/5(z + 2) - (4/25) + (9/125)(z + 2) - (12/625)(z + 2)^2 \dots$

4.16 (a) $\sum_{n=0}^{\infty} z^{-n}/n!$

  (b) $\sum_{n=0}^{\infty} (-1)^n z^{-2n}/n!$

4.17 $\sum_{n=0}^{\infty} (-1)^n z^{-2n-1}/(2n+1)!$; r can be arbitrarily large.

4.18 $(z - 1)^2 + (z - 1)^{-2}$.

4.21 (a) zero of order 1 at $z = 0$; pole of order 1 at $z = -1$.
  (b) no zeros; simple poles at $z = \pm i$.

(c) simple zeros at $z = 2k\pi i$, k any integer; no poles.

(d) no zeros; simple poles at $z = 2k\pi i$, k any integer.

(e) simple zeros at $z = \exp(2k\pi i/3)$, $k = 0,1,2$; simple poles at $z = \text{Log } 2 + 2m\pi i$, m any integer.

(f) zeros of order 2 at $z = \pm 1$; pole of order 2 at $z = 0$.

4.23 Yes.

4.24 $z = (2k + 1)\pi/2$; k any integer; order 1.

4.25 pole of order 2 at $z = 0$; pole of order 1 at $z = k\pi$ for each nonzero integer k.

4.28 (b) -1.

Chapter 5

5.1 (a) 1

(b) $1/4$ at $z = \pm 1$, $-1/4$ at $z = \pm i$

(c) $(-1)^k$ at $z = k\pi$, k any integer.

(d) $-1/\pi$ at $z = (2k + 1)/2$, k any integer.

(e) $e/2$ at $z = 1$, $-e^2$ at $z = 2$, $e^3/2$ at $z = 3$.

(f) 1 at $z = 0$.

(g) $1/2$ at $z = 0$

(h) -1 at $z = 0$, $e(1 - \sin 1 \cos 1)/\sin^3 1$ at $z = 1$.

5.2 (a) πi.

(b) 0

(c) 0

(d) $-\pi i$

(e) $2\pi i$

(f) 0

5.4 $6\pi i$

5.6 $m_1 + m_2 + \ldots + m_n$

5.7 $-10/\pi$

5.13 (a) $2\pi/3^{1/2}$

(b) $\pi/32$

(c) $2\pi \exp(-3^{1/2})\cos 1/3^{1/2}$

(d) $\pi \exp(-\alpha/2^{1/2}) \sin(\alpha/2^{1/2})$

5.14 $(\pi/8)(1 + e^{-4})$

5.15 $\pi/2$ if $m > 0$, $-\pi/2$ if $m < 0$, 0 if $m = 0$.

5.23 $\pi/3^{1/2}$

5.24 $(-\pi/10)(e^{-1} + \sin 2/2)$

5.25 $\beta > 9/2$

5.27 all of them.

5.31 $(q_1b_1 + q_2b_2 + q_3b_3) - (p_1a_1 + p_2a_2 + p_3a_3)$

5.32 (a) $-2\pi i$
     (b) $-4i$

5.35 (a) $[(2\pi)^{1/2}/4] \exp(-2|y|)$
     (b) $[(2\pi)^{1/2}/4](1 + |y|) \exp(-|y|)$
     (c) $(2\pi/3)^{1/2} \exp(-|y|\sqrt{3}/2 - iy/2)$
     (d) $(2\pi)^{-1/2}[1/(a + iy)]$
     (e) $(2\pi)^{-1/2}[4 \sin^2(y/2)/y^2]$, if $y \neq 0$; $(2\pi)^{-1/2}$, if $y = 0$

5.36 In all cases $\mathfrak{F}^{-1}[\mathfrak{F}(f)] = f(x)$.

5.37 (a) $k(2/\pi)^{1/2}(\sin y/y)$, if $y \neq 0$; $k(2/\pi)^{1/2}$, if $y = 0$.
     (b) $k^2(2\pi)^{-1/2}(2 - |x|)$, if $|x| < 2$; 0, if $|x| \geq 2$.
     (d) Yes.

5.38 (a) $(2/\pi)^{1/2} \exp(-iy/2)\sin(y/2)/y$, if $y \neq 0$; $(2\pi)^{-1/2}$, if $y = 0$.
     (b) $(2/\pi)^{1/2}/(1 + y^2)$, if $y \neq 0$; $(2/\pi)^{1/2}$, if $y = 0$.
     (c) $(2/\pi)^{1/2}/[1 + (y - 2)^2]$, if $y \neq 2$; $(2/\pi)^{1/2}$, if $y = 2$.

5.43 $(1/k(\pi t)^{1/2})\exp(-x^2/4k^2t)$

5.44 (b) $U(t,y) = \mathfrak{F}(f) \cos(kyt) + [\mathfrak{F}(g)/ky] \sin (kyt)$
     (c) $(1/2)\{1/[1 + (kt + x)^2] + 1/[1 + (kt - x)^2]\}$
         $+ (1/2k)[\arctan(x + kt) - \arctan(x - kt)]$

Chapter 6

6.1 (a) $\Delta = \{w: e^{-1} < |w| < e, -\pi/2 < \arg w < \pi/2\}$

6.2 The image of D is the region
     $\{(u + iv): u > 0, v > 0, (u/\cosh 1)^2 + (v/\sinh 1)^2 = 1\}$.

6.7 (a) $\exp(i\pi/2)[(z - 1/2)/(1 - z/2)]$
     (b) $2i(1 + z)/(1 - z)$
     (c) $-iz$
     (d) $(z^4 - i)/(z^4 + i)$
     (e) $\sin(z/2 + \pi/4)$
     (f) $\exp[(1 + i)z/\pi]$
     (g) $2/(2 - z)$

6.10 $f(z,\lambda) = A + BM \int \dfrac{dz}{(z + 1)^{\alpha_1/\pi} z^{\alpha_2/\pi} (z - 1)^{\alpha_3/\pi}}$, where A, BM are constants, and α_1, α_2, α_3 depend on λ. In the limit $f(z) = (1/2)\log(z^2 - 1)$.

6.13 (a) $(1/\pi)\text{Arg}[(z^2 - 1)/(z^2 + 1)]$
(b) $(4/\pi)\text{Arg } z$
(c) $(3/\pi)[\text{Arg}(e^{\pi z} + 1) + \pi - \text{Arg}(e^{\pi z} - 1)]$
(d) $(1/\pi)[\text{Arg}(\sin z + 1) + \pi - \text{Arg}(\sin z - 1)]$
(e) $(1/\pi)\{\text{Arg}[(z + 1)^2/2] - \text{Arg}[(z - 1)^2/(z + 1)^2]$
$+ \pi - \text{Arg}[(z - 1)^2]\}$
(f) $(1/\pi)\text{Arg}[i(1 + z)/(1 - z)]$

6.14 $k_1/4 + 3k_2/4$

6.15 $e^{-y}\cos x$

6.16 $(2/\pi)\text{Arg}(e^z\} - 1$

6.17 $1 + (1/\pi)\text{Arg}(z + 1/z)$

6.19 $g(z) = A + (B - A)\left[\dfrac{\text{Log } |w| - \text{Log } a}{\text{Log } b - \text{Log } a}\right]$, where
$w = (z - 3^{1/2})/3^{1/2}z - 1)$.

6.20 $- \log |(z - i\alpha)/(z + i\alpha)|$

6.21 $(4/\pi)\text{Arg } z - 2 + 2 \log |(z - i)/(z + i)|$

6.22 center approaches i and radius approaches 0.

6.23 $- \log |z/2 + 2/z|$

6.24 $(2/\pi)\text{Arg } z - 1 - 3 \log|z|$

6.25 $2 - (4/\pi)\text{Arg}[i(2 - i + z)/(2 + i - z)] + \log|(z - i)/2|$
$- \log|2z/(5 + iz)|$

6.26 $(2/\pi)\text{Arg}[(z + 1)^2/(z - 1)^2] -$
$(1/2)\log\left[\dfrac{25(1 + z)^2 + (7 - 24i)(1 - z)^2}{25(1 - z)^2 + (7 + 24i)(1 + z)^2}\right]$

Chapter 7

7.1 $H(w) = k(w + 1/w)$, $F(w) = k(1 - 1/\overline{w}^2)$; equipotentials $u + u/(u^2 + v^2) = $ constant, streamlines $v - v/(u^2 + v^2) = $ constant.

7.2 $H(w) = k(1 + e^{-2w})^{1/2}$,
     $F(w) = -[ke^{-2w}/(1 + e^{-2w})^{1/2}]$

7.3 (a) $-(1/2)\log\{[s^2 + r^2 - 2rs \cos(\phi - \theta)]/$
         $[1 + r^2s^2 - 2rs \cos(\phi - \theta)]\}$

Chapter 8

8.1 (a) $w^{1/2}$
     (b) $\pm i$

8.2 (a) $f'(z) = 2e^z$
     (b) $f(z) = 2e^z - 1$

8.3 (a) $f'(z) = g'(z) + g(z)$, where $f(0) = g(0)$.

8.5 $\pi/2$

8.6 $G(z) = F'(z)$.

8.7 (a) for all z except $\{z : z$ is real, $|z| \geq 1\}$
     (b) $2\pi(1 - x^2)^{-1/2}$, $-1 < x < 1$.
     (c) $2\pi(1 - z^2)^{-1/2}$

8.9 $\Gamma(z)\Gamma(1 - z) = \pi/\sin \pi z$

Index